Bruna Medeiros Pereira

Analysing the overexpression of candidate genes for resistance

Bruna Medeiros Pereira

Analysing the overexpression of candidate genes for resistance

Meloidogyne spp. in transgenic peanut and soya roots

ScienciaScripts

Imprint
Any brand names and product names mentioned in this book are subject to trademark, brand or patent protection and are trademarks or registered trademarks of their respective holders. The use of brand names, product names, common names, trade names, product descriptions etc. even without a particular marking in this work is in no way to be construed to mean that such names may be regarded as unrestricted in respect of trademark and brand protection legislation and could thus be used by anyone.

Cover image: www.ingimage.com

This book is a translation from the original published under ISBN 978-613-9-64058-4.

Publisher:
Sciencia Scripts
is a trademark of
Dodo Books Indian Ocean Ltd. and OmniScriptum S.R.L publishing group

120 High Road, East Finchley, London, N2 9ED, United Kingdom
Str. Armeneasca 28/1, office 1, Chisinau MD-2012, Republic of Moldova, Europe
Printed at: see last page
ISBN: 978-620-7-72007-1

ACKNOWLEDGEMENT

The Coordination for the Improvement of Higher Education Personnel (CAPES) for making the scholarship available.

Embrapa Recursos Genéticos e Biotecnologia for providing the infrastructure that made it possible to carry out the proposed project.

Dr Regina Carneiro, for providing essential material and infrastructure at all stages involving nematodes during the project, for her support and guidance.

Dr Ana Cristina Miranda Brasileiro for always being there whenever I needed her, with the best solutions for any obstacle I encountered. For her trust, for making the project available, for her guidance, advice and encouragement at every stage of my postgraduate studies.

To Dr Larissa Arrais, who was essential in the first year of my project, thank you for teaching me rigorously, for making me see beyond the immense beauty of research, for being so competent and dedicated.

To my dear friends at the Plant-Pest Molecular Interaction Laboratory III (LPPIII) at Embrapa Genetic Resources and Biotechnology, who were there for me in good times and bad, always supporting me. In particular, my friends Andressa Martins, Thaís Nicolini and Eliza Bellard, for being there every day, giving me strength and the will to carry on, bringing great joy to my routine and helping me whenever I needed it.

To my family and friends (Renata Bomfim, Carolina Araújo, Aline Rigotti, Izabela Pinheiro, Joana Furtado, Raquel Bomfim and Aline Barbosa) who, even though they come from other fields of study, have always shown interest and overflow with unconditional love and support that keeps me going at every stage of my life.

SUMMARY

SUMMARY

Cultivated peanuts *(Arachis hypogaea* L.) have a high nutritional quality and are characterised as a rich source of energy and amino acids. Its productivity is limited by several factors, including its susceptibility to the gall nematode *(Meloidogyne arenariã).*

On the other hand, the wild relatives of the peanut *(Arachis* spp.) show great genetic variability and are a potential source of resistance alleles for breeding programmes. Transcriptome studies of the interaction of *M. areanaria* with resistant wild peanut species, such as *A. duranensis* and *A. stenosperma,* have indicated some candidate genes potentially involved in resistance to attack by this pathogen.

From these studies, several candidate genes were selected and their differential expression during the interaction was subsequently validated. The aim of this work is to understand the biological role of the candidate genes *AdEXLB8, AsDUF* and *AsLIP,* from wild species, in resistance to the root-knot nematode *(Meloidogyne* spp), through their overexpression in transgenic roots of peanut and soya *(Glycine max),* susceptible species.

To do this, the ORFs (Open Reading Frame) of these genes were cloned into the binary vector pPZP-201BK-EGFP under the control of the actin promoter and introduced into the petiole of peanut leaves or into the hypocotyl of soya by means of transformation with a wild strain of *Agrobacterium rhizogenes.* Transgenic peanut and soya roots were subsequently inoculated with *M. arenaria and M. incognita,* respectively, and evaluated 60 days after inoculation for disease development. Roots overexpressing each of the three candidate genes showed, in both peanut and soya, a significant reduction in the number of galls and egg formation when compared to roots transformed only with the empty vector. *In planta* functional analysis of the phenotypic effects of overexpressing *AdEXLB8, AsDUF* and *AsLIP in* transgenic peanut and soya roots infected with different species of *Meloidogyne* indicate an involvement of these genes in resistance to the gall nematode.

Keywords: *Arachis hypogaea, Glycine max,* genetic transformation, *Meloidogyne* spp., resistance genes.

CHAPTER 1

INTRODUCTION

The genus *Arachis* (Fabaceae family) is native to South America, with a probable centre of origin in the higher altitude regions of Central Brazil (Gregory et al., 1980), and a natural distribution restricted to Brazil, Bolivia, Paraguay, Argentina and Uruguay (Valls and Simpson, 1994). The peanut *(A. hypogaea), the* only species of the *Arachis* genus cultivated on a large scale, has seeds rich in oil and protein, which are produced below the surface of the soil and can be eaten both *fresh* and processed. The peanut is a leguminous oilseed of great economic importance, widely consumed in Asia, Africa and the Americas, and shows great potential in combating hunger and malnutrition in underdeveloped countries. Peanut production is severely affected by various phytopathogens (Vargas Gil et al., 2008), such as leaf fungi and nematodes. The most damaging and economically important phytoparasitic nematodes for groundnuts are the gall-forming nematodes *(Meloidogyne* spp.), in particular *M. arenaria* (Mitreva et al., 2005).

The use of resistant cultivars is important in the fight against gall-forming nematodes. *Mi* genes have been successfully used in the development of cultivars resistant to gall-forming nematodes, as the presence of these genes reduces the infection of different species of *Meloidogyne* (Charchar et al., 2004). However, as a probable result of the continuous field exposure of plants resistant to gall nematodes (Tzortzakakis and Gowen, 1996), some nematode species have developed the ability to break down the resistance conferred by *Mi* genes, making it necessary to continue searching for new sources of resistance to *Meloidogyne* spp. populations (Charchar et al., 2004). The emergence of virulent nematode populations in previously resistant cultivars is cause for concern, as it affects the durability of the resistance genes used (Semblat et al., 2000).

Despite its economic importance, the cultivated peanut has little genetic variability, while its wild relatives have a broader genetic base and a large number of resistance-related alleles to be exploited. Embrapa Recursos Genéticos e Biotecnologia has an active germplasm bank (BAG) with more than 1,300 accessions of 79 wild *Arachis* species, some of which have been identified as a source of resistance to fungi and nematodes (Leal-Bertioliet al., 1999; Bertioli et al., 2000; Fávero 2004). Studies aimed at exploiting this genetic pool have shown that *A. stenosperma,* a wild relative of peanut, is resistant to the gall nematode *M. arenaria* by inducing a hypersensitivity response (HR) at the site of nematode penetration (Proite et al., 2008). Subsequently, sequencing of the *A. stenosperma* transcriptome and *in vitro* analysis of gene expression enabled the identification of a large number of candidate genes involved in the HR to the gall nematode (Proite et al., 2007; Guimaraes et al., 2010; Morgante et al., 2013; Guimaraes et al., 2015). In the present work, three of these candidate genes *{AsDUF, AsLIP* and *AdEXLB8)* were selected to study the effects of their overexpression on the response to attack by gall-forming nematodes *{M. arenaria* and *M. incognita)* in two susceptible species of peanut and soya *{Glycine max).*

The *AsDUF* gene, identified in *A. stenosperma,* encodes a protein with unknown function that has a conserved domain called *Domain of Unkovm Function* (DUF). *AsLIP* encodes a protein called Lipocalin in *A. stenosperma* that is involved in regulating salicylic acid levels, which often lead to HR and are involved in modulating tolerance to oxidative stress (Zinovieva et al., 2011). *AdEXLB8* encodes an Expansin-ZzAe B from *A. duranensis* and is involved in cell wall plasticity in response to environmental stimuli such as phytopathogen attack (McQueen-Mason et al.,1992; Marowa et al.,2016). The three genes *{AsDUF, AsLIP* and *AdEXLB8)* selected for this study are positively regulated in response to *M. arenaria* infection in *A. stenosperma* previously described by (Morgante et al., 2013).

Identifying the biological function of candidate genes of potential agronomic interest that have not yet been characterised is of fundamental importance in order to add innovation to the information generated by plant functional genomics projects. In the process of studying and evaluating the biotechnological potential of these candidate genes, it is essential to analyse their expression *in planta.* However, analysing expression in the target plant can be complex, time-consuming and expensive, particularly when the plant is recalcitrant to genetic transformation, such as peanuts. In order to overcome this limitation, this study developed methods for inducing transgenic roots in peanut and soya composite plants that allow the efficient and rapid study of the effects of overexpression of candidate genes (Bevan 1984; Simpson et al., 1986). With these methods, transgenic roots are obtained by inoculating the plant with wild strains of *Agrobacterium rhizogenes,* containing a binary vector for overexpression in plants of the candidate genes. Composite plants are thus obtained, in which the aerial part is non-transgenic and its original roots replaced by transgenic roots overexpressing the candidate genes.

CHAPTER 2

OBJECTIVES

2.1.1 ubjective

To study the effects of overexpressing three candidate genes *{AsDUF, AsLIP* and *AdEXLB8)* in transgenic peanut *{Arachis hypogaea)* and soya *{Glycine max)* roots and to evaluate their potential as candidates for conferring resistance to attack by gall-forming nematodes *{Meloidogyne* spp.)

2.1.2 bjectives

- To establish a methodology for obtaining transgenic roots from alternative explants (detached leaves) of peanut (*A. hypogaea)* and for inoculating them with the gall nematode *(M. arenaria);*
- To establish a methodology for obtaining composite soya plants *{Glycine max)* and for inoculating them with the gall nematode *{M. incognita);*
- Transform peanut roots with three candidate genes *{AsDUF, AsLIP* and *AdEXLB8)* and select transgenic events;
- Transforming soya roots with candidate genes that show the best results in peanuts and selecting transgenic events;
- To evaluate the effects of overexpression of candidate genes on the development and reproduction of *Meloidogyne* spp. in transgenic peanut and soya roots.

CHAPTER 3
LITERATURE REVIEW

3.1. Growing peanuts

The *Arachis* genus has its centre of origin in South America, more specifically in the Serra de Amambai, on the border between Mato Grosso do Sul and Paraguay, where the oldest species of the genus, *A. guaranitica,* occurs (Gregory et al., 1980). The spread of the cultivated peanut *(A. hypogaea)* began by the indigenous people to the various regions of South America, Central America and Mexico. It was later introduced to Europe. In the 19th century, peanuts spread from Brazil to Africa and from Peru to the Philippines, China, Japan and India (Hammons, 1982 cited in Wynne et al., 1991).

Currently, peanuts are the fourth largest oilseed crop in the world, behind soya, canola and cotton (USDA-FAS, 2015). Its importance among oilseeds is due to its high nutritional quality, with 44 to 56 per cent oil and 22 to 30 per cent protein, depending on the variety and environmental conditions. Peanuts also play a role in animal nutrition, and can be used directly as pasture, in the case of forage peanuts *(A. pintoi),* or the by-products of oil extraction, such as cake and bran, used as animal feed (Evangelista et al., 2004). In addition, peanuts contribute to soil fertility as they have a root system capable of fixing nitrogen and can also be used in crop rotation or succession with other species such as sugar cane and maize.

In Brazil, due to the expansion of soya cultivation, driven by the oil industry and the needs imposed by the world market, peanut cultivation areas and production have gradually decreased, stabilising in the range of 90,000 to 100,000 hectares cultivated, with yields of between 171,000 and 217,000 tonnes/year between 1998 and 2003, and 288,000 to 301,000 tonnes/year between 2004 and 2006 (Conab, 2004 and 2006). Since 2000, Brazil has expanded peanut production in the South and Southeast, thanks to significant increases in productivity due to the introduction of new cultivars. The most widely used peanut varieties in Brazil originate from the Agronomic Institute of

Campinas, which carries out research and seed development activities (Conab, 2004).

3.2. Galling nematodes on groundnuts

In the world's most diverse crops, phytoparasitic nematodes have been responsible for a significant proportion of the losses caused by their harmful action on the root system, affecting the absorption and translocation of nutrients and altering the plant's physiology. Species of the genus *Meloidogyne,* known as gall-forming nematodes, stand out among the main pathogens and represent a threat to agricultural production on a global scale, causing losses of up to 100 million dollars a year (Ibrahim et al., 2011). In countries such as the United States and India, they cause up to 12% of production losses (Bailey, 2002).

M. arenaria race 1, M. *hapla, M. javanica* and *M. haplanaria* (Eisenbacket al., 2003) limit the development of peanut plants worldwide, causing symptoms such as wilting, loss of colour, rotting of gynophores, pods and roots. The association of these nematodes with soil fungi causes losses of US$ 5 billion per year in the United States (Williamson and Hussey, 1996).

Galls nematodes are sedentary and obligate endoparasites, feeding exclusively in the cytoplasm of plant cells (Williamson and Gleason, 2003). All nematodes of the genus *Meloidogyne* have parasitism characterised by the establishment of permanent feeding sites that include giant cells in the cortex, endodermis, pericycle and the vascular parenchyma of the host plant root (Trudgill and Blok, 2001). The feeding sites are like drains of nutrients produced through photosynthesis by the plant, causing deformations and blocking the vascular tissues, obstructing the translocation of water and nutrients, thus limiting the growth and development of the plant. The root tissues surrounding the giant cells form swollen structures called galls, which are the main symptom used to assess the degree of infection (Gheysen and Fenoll, 2002). The galls formed by *M. arenaria* on peanuts can grow into swellings twice the diameter of the root (Williamson and Hussey, 1996).

Control strategies for gall nematodes on groundnuts include crop rotation or succession with non-host or antagonist species such as Crotalaria spp. (crotalaria), *Mucuna spp. (*mucunas) and *Tagetes* spp. (cloves) (Huang et al., 1980; Huang and Charchar, 1981) and the use of nematicides. Unfortunately, these practices have little effect on another nematode that peanuts are also host to, *Pratylenchus* spp. and chemical control is inefficient and poses risks to human health and the environment. Therefore, the development of peanut varieties resistant to the gall nematode is a strategy of great interest for its control.

3.3. Wild species of *Arachis*

The *Arachis* genus comprises around 80 species and is divided into nine botanical sections according to morphology, geographical distribution and crossbreeding viability (Krapovickas and Gregory, 1994; Valls and Simpson, 2005). The cultivated peanut, *A. hypogaea,* is an allotetraploid species (2n = 4x = 40), composed of two distinct genomes (genomes A and B), which arose from an eventual cross between the two diploid wild species, *A. duranensis* and *A. ipaènsis.* The result of the cross was a sterile hybrid, which showed a doubling of the number of chromosomes by a natural event, causing fertility to be restored (Kocher et al., 1996; Lavia et al., 2001; Jung et al., 2003; Seijo et al., 2004).

Due to allotetraploidisation, the cultivated peanut has remained reproductively isolated from its wild relatives, thus narrowing its genetic base. In contrast, the wild diploid species of *Arachis* spp. are genetically more diverse, making them a potential source of alleles for improving the cultivated species (Hilu and Stalker, 1995). For example, the wild species *A. stenosperma* has high levels of resistance to some phytopathogens, including the gall nematode *M. arenaria, through the* manifestation of the Hypersensitivity Response (HR) (Proite et al., 2007). In this species, the induction of feeding sites by the nematode was associated with a necrotic response, hindering the development of the nematode, the formation of galls and egg masses (Proite et al., 2008). Studies involving cDNA libraries, macroarray, mass sequencing and expression analysis by quantitative PCR in *A. stenosperma* and other wild peanut species, such as *A. duranensis,* have indicated some candidate genes potentially involved in resistance to attack by this pathogen (Proite et al., 2007; Guimaraes et al., 2010; 2015; Morgante et al., 2013). Some of these candidate genes are involved in the salicylic acid (NBS-LRR, lipocalins, resveratrol synthase) and jasmonic acid (patatin, allene oxide cyclase) pathways and are also related to hormonal balance (auxin responsive protein - AUX/IAA and glycosyl hydrolase 3 - GH3), cell plasticity and cell signalling (tetraspanin, integrin, expansin) (Guimaraes et al., 2015). However, to date, the mechanisms involved in the HR response of *A. stenosperma* to *M. areanaria have* not been fully elucidated.

3.4. *Arachis* spp. candidate genes for gall nematode tolerance

One of the candidate genes from the aforementioned studies is called *AdEXLB8*, which encodes an Expansin-like B (EXLB) belonging to the Expansin superfamily that is related to cell wall plasticity, regulating plant expansion and growth (Cosgrove, 2015). Previous studies have shown a positive regulation of *AdEXLB in the in silico* analysis of the transcriptome of *A. stenosperma* roots inoculated with *M. arenaria in* the first stages (3, 6 and 9 days) after inoculation (Guimaraes et al., 2015). In parallel, this gene was also shown to be positively regulated in the transcriptome of *A. duranensis* roots during the process of gradual water deficit (Brasileiro et al., 2015). Subsequent analyses of this gene using the RT-qPCR technique, both after inoculation with *M. arenaria* and during gradual dehydration, made it a potential candidate for simultaneously conferring resistance to *M. arenaria* and drought tolerance (Guimaraes et al., 2015; Brasileiro et al., 2015). Plant Expansins are related to responses to abiotic and biotic stresses, such as drought, heat, cold and infection by phytopathogens (Marowa et al., 2016), indicating that these proteins constitute a common component in plant-stress interaction. Expansins are also involved in oxidative stress responses (Han et al., 2015), derived from the production and accumulation of reactive oxygen species (ROS) (Baxter et al., 2014).

Another candidate gene chosen for this study is *AsDUF,* which encodes a protein that contains a conserved DUF *(Domain of Unknown Function)* domain belonging to a family of unknown functions (Gholizadehet al., 2011). These proteins are found in response to various types of biotic stresses, such as attacks by bacteria and fungi, and abiotic stresses, such as drought and salinity (Gholizadehet al., 2011). The DUF family of proteins has already been identified in *Arabidopsis,* rice and tomato plants under different biotic and abiotic stress conditions and may be related to plant defence (Brunings et al., 2009; Gholizadeh et al., 2011). A DUF protein isolated from *Celosia cristata* showed an increase in the activity of enzymes that induce HR, such as catalase, peroxidase, polyphenol oxidase and phenylalanine ammonia lyase, when applied to tobacco leaves (Gholizadeh et al., 2011). The *AsDUF* gene was selected as a candidate for the present study due to recent results showing an increase in its expression at 3, 6 and 9 days after inoculation with *M. arenaria* in *A. stenosperma* plants by means of RT-qPCR analyses (Morgante et al., 2013).

Another candidate gene from *Arachis* transcriptome studies is the one that encodes a Lipocalin involved in regulating salicylic acid levels, which often lead to RH, and in modulating tolerance to oxidative stress
(Zinovieva et al., 2011). Lipocalins have been shown to play an important role in the removal of potentially harmful molecules, ROS, which are chemical compounds resulting from the activation or reduction of molecular oxygen or derived from reduction products, known to be induced by stresses related to extreme temperatures and excess light (Charron et al.,2008). Recently, the *AsLIP* gene was identified as being positively regulated *in in silico* analysis of the transcriptome of *A. stenosperma* roots inoculated with *M. arenaria in* the early stages (3, 6 and 9 days) after inoculation (Guimaraes et al., 2015). RT-qPCR analyses validated the expression profile of this gene (Morgante et al., 2013), making it a potential candidate for resistance to *M. arenaria* in transgenic peanut plants.

3.5. Expression of candidate genes in transgenic peanut roots

The process of studying and analysing the biotechnological potential of candidate genes is of great importance to agriculture. The information generated by plant functional genomics projects, together with *in planta* analysis of the expression of these genes, is essential in the search for new sources of resistance. Isolating genes and transferring them to economically important cultivated species, such as peanuts, through the genetic transformation of plants is a promising strategy. However, peanuts *(A. hypogaea) are* recalcitrant to genetic transformation, making this process difficult.

Therefore, among the different genetic transformation systems already tested in peanuts, the method selected for the evaluation of candidate genes for nematode resistance was that of obtaining a composite plant through transformation mediated by *Agrobacterium rhizogenes,* as it is efficient and generates results quickly. This method allows us to obtain composite plants, where only part of the plant is transformed, bypassing the difficulty of stable genetic transformation.

A. rhizogenes is a gram-negative soil bacterium that was first identified more than 70 years ago (Riker et al., 1930; Hildelbrand, 1934; White, 1972) as the causal agent of hairy root syndrome. *A. rhizogenes* infects dicotyledons, inducing the proliferation of roots at the site of infection and in the process transferring DNA fragments from the bacterium to the roots. Transformation mediated by wild strains of *A. rhizogenes,* where transgenic roots are obtained by inoculating the plant with strains of this bacterium, is of great importance due to the species' ability to transfer DNA fragments to plant cells without damaging the plant's cell wall, using T-DNA (transfer DNA). In this way, a fragment containing the candidate gene is introduced into the plant and expressed in transgenic roots, resulting in composite plants where the aerial part is non-transgenic and its roots are replaced by transgenic roots expressing the candidate genes.

This is a fast and efficient transformation system that allows the simultaneous analysis of a large number of candidate genes in various plant species, including *Glycine max* (Li et al., 2010), *Phaseolus vulgaris* (Estrada-Navarrete et al., 2006), *Prunus* spp. (Bosselut et al., 2011), *Cqffea arabica (*Alpizar et al., 2006), *Medicago truncatula (*Deng et al., 2011), *Arachis hypogaea* (Liu et al., 2016). The simplicity and efficiency of this transformation method favours its application in studies of signal transduction, root nodule formation, plant-bacteria interaction and the study of genes that confer nematode resistance (Kosuta et al., 2008; Cai et al., 1997; Tirichine et al., 2007; Lefebvre et al., 2010). In the composite plant transformation system, each transgenic root represents an independent transformation event (Keresztet al., 2007) and multiple transgenic roots can be recovered from a single plant or explant.

Different methodologies can be used to obtain composite plants. Specifically in the peanut crop, Geng et al. (2012) obtained transgenic roots in 45 days by transforming embryonic axes with *A. rhizogenes,* with an efficiency rate of up to 61%. Liu et al. (2016) used the transformation strategy with *A. rhizogenes* in the petiole region of detached peanut leaves, and the efficiency of transformed roots was over 90%.

In the composite plant transformation system each transgenic root represents an independent transformation event (Keresztet al., 2007), multiple transgenic roots can be recovered from a single plant or explant, making *A. rhizogenes-mediated* transformation an important tool for analysing candidate genes.

CHAPTER 4

MATERIAL AND METHODS

4.1 Where the experiments were carried out

The experiments were carried out at the Plant-Pest Interaction Laboratory III (LPP-III) of Embrapa Genetic Resources and Biotechnology (CENARGEN) located in Brasilia-DF, Brazil.

4.2 Cloning the three candidate genes

The DNA sequences corresponding to the ORFs (Open Reading Frame) of the three candidate genes targeted in this study, i.e. *AdEXLB8, AsDUF* and *AsLIP,* were synthesised and cloned into the binary plant expression vector pPZP_201BK_EGFP (Figure 1) by the company Epoch Life Science (Texas, USA) (Chuet al., 2014). Each of the ORFs of the *AdEXLB8, AsDUF* and *AsLIP* genes was cloned into the unique *Xho* I restriction site present in the expression cassette of the pPZP_201BK_EGFP vector for overexpression in plants, under the control of the Actin 2 (ACT-2) promoter from *Arabidopsis thaliana* (Yang et al., 2003) and the Nopaline Synthase (NOS) terminator from *Agrobacterium tumefaciens* (Novak et al., 2000). The binary vector pPZP_201BK_EGFP, from now on referred to only as pPZP, also has two other cassettes between the edges of its T-DNA: one with the *enhanced Green Fluorescent Protein (eGFP)* reporter gene, directed by the 35S promoter of the cauliflower mosaic virus (35SCaMV) and NOS terminator (Novak et al., 2000) and another with the hygromycin phosphotransferase *(hpf)* selection marker gene from *Escherichia coli,* which confers resistance to the antibiotic hygromycin, directed by the Ubiquitin (UBI3) promoter (Hajdukiewicz et al., 1994) from *A. thaliana* and the NOS terminator.

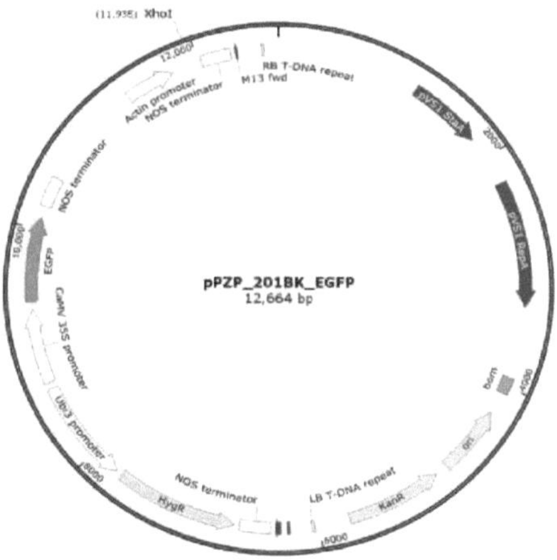

Figure 1 - Map of the pZP_201BK_EGFP vector (Chu et aL, 2014) which contains the following cassettes between the T-DNA borders: cassette for overexpression in plants of candidate genes under the control of the Actin 2 (ACT-2) promoter and the Nopalin Synthase (NOS) terminator to be cloned into the unique *Xhol'* site, cassette with the *enhanced* reporter gene. *green fluorescent protein (eGFP) reporter gene* directed by the 35S promoter and NOS terminator, and the third with the hygromycin phosphotransferase *(hpt)* selection marker gene directed by the Ubiquitin (UBI3) promoter and NOS terminator.

4.3 Transformation of *Agrobacterium rhizogenes* strain K599 with the binary vectors

4.3.1 Electroporation

The wild strain K599 of *A. rhizogenes,* kindly provided by Dr Peggy Ozias-Akins (University of Georgia, USA) was transformed with the pPZP vector (empty vector) and with pPZP containing the candidate genes *(AdEXLB8, AsDUF and AsLIP), by* means of electroporation, according to Brasileiro and Cameiro (2015). In summary, 1 µL of plasmid DNA (10 pg) was added to a 1.5 mL microcentrifuge tube containing approximately 40 µL of *A. rhizogenes* K599 suspension in 10% glycerol (v/v), mixed gently and incubated in the

ice for about two minutes. The suspension was transferred to an electroporation cuvette cooled to 4°C and electroporated at 1.8 kV in the MicroPulser electroporator (BioRad, California, USA), with a capacitance of 25μF and a resistance of 400 Ω. Immediately after the pulse, 1 mL of liquid LB medium (Miller, 1972) was placed in the cuvette and mixed with the suspension using a micropipette. Again, this suspension was transferred to a 1.5 mL microcentrifuge tube and incubated at 28°C for two hours. After this period, different volumes (50, 100 and 200 μL) of the suspension were distributed in Petri dishes containing semi-solid LB medium with the selection agents kanamycin (100 mg/mL) and streptomycin (100 mg/mL) and incubated for two days at 28°C for 48 h to grow bacterial colonies.

4.2.1 Selection of transformed *A. rhizogenes* strains

After two days of incubation, six isolated colonies grown on selective medium were randomly selected to confirm the presence of the pPZP vector isolated or containing the candidate genes *AdEXLB8, AsDUF* and *AsLIP.* To confirm transformation, PCR (Polymerase Chain Reaction) was carried out on 1 μL of the cell suspension of isolated colonies diluted in 20μL of MilliQ H2O using specific *primers.* The DNA from each colony thus obtained, was used as a template for a PCR reaction using a pair of *primers* that amplify the *eGFP* gene *(eGPP_Forward5'~* CGTCGCCGTCCAGCTCGACCAG-3 ' and eGFPJteverae5'-CATGGTCCTGCTGGAGTTCGTG-3') and another that amplifies the flanking region of the candidate gene (pPZP_Forvran75'-CTACCAGAATTTGGCTTGAC-3' and pPZP_ *Reverse* 5'-GTAAAACGACGGCCAGT-3'). The PCR reaction was carried out using the Taq DNA Polymerase Recombinant enzyme (Invitrogen, California, USA) in a final volume of 25 μL, according to the manufacturer's instructions. The cycle used in the Master Cycler thermal cycler (Eppendorf, California, USA) was: 5 minutes at 94 C°; 35 cycles of 30 seconds at 94 C°, 30 seconds at 50 C°, 30 seconds at 72 C°; and 7 minutes at 72 C°.

The size and quality of the amplification product *(amplicori)* for each colony was checked using 1.5% (w/v) agarose gel electrophoresis. After confirmation by PCR, a positive colony for each construct (pPZP; pPZP- *AdEXLB8, pPZP-AsDUF and* pPZP-AsLIP*)* was inoculated into 3 mL of liquid LB medium containing kanamycin

(100 mg/mL) and streptomycin (100 mg/mL) for 16 hours at 28 °C. An aliquot of this culture medium was stored in 15% (v/v) glycerol in a -80 °C freezer for later use.

4.4. Multiplication of *Meloidogyne* spp. in tomato plants

To inoculate and multiply the gall nematode *(Meloidogyne* spp.) in tomato plants *(Solanum lycopersicum* L.) of the Santa Clara variety, soil previously infected with *M. arenaria* race 1 was used, kindly provided by Dr Regina Carneiro (Nematology Laboratory of Embrapa Recursos Genéticos e Biotecnologia, Brasília-DF). After 60 days of cultivation in a greenhouse in the presence of the nematode, the root system of the tomato plants was collected to extract the eggs, according to the methodology described by Boneti and Ferraz (1981). In short, the root system of each individual was cut into pieces of approximately 2 cm and ground in a blender in the presence of a 1% (v/v) sodium hypochlorite solution (NaOCl) for 40 seconds. The suspension was then passed through a set of sieves with openings of 0.297, 0.149 and 0.025 mm, arranged in that order. The eggs were recovered from the 0.025 mm sieve and transferred to a modified Baerman funnel and kept at room temperature. Individuals at juvenile development stage 2 (J2) were obtained after the eggs hatched, collected every two days and stored at 4 °C for eight days until the end of the collections. After three collections, the J2 individuals were quantified on a special slide called a Peters chamber, which contains two overlapping slides with a space or chamber between them, where 1 mL of the nematode suspension was placed for counting under an optical microscope (AxiosKop, Zeiss - Oberkochen, Germany). The individuals were then resuspended in distilled water at a concentration of 1,000 individuals/mL.

4.5 Transformation of detached groundnut leaves with *A. rhizogenes*

4.5.1 Obtaining detached leaves

Fully expanded young leaves containing four leaflets were detached from peanut plants *(A. hypogaea* 'Runner IAC-866') grown in the greenhouse two months after germination, leaving approximately 5 cm of petiole. The detached leaves were then placed in Petri dishes (100 x 15 mm) with a layer of cotton wool at the bottom, a sheet of filter paper on top and a glass slide to support the leaflets (Figure 2), using the detached leaf method (Moraes and Salgado, 1982) adapted for compound transformation. To keep the humidity high in the Petri dish, the cotton wool and filter paper were soaked daily with distilled water. A total of 60 Petri dishes were prepared, each containing a detached peanut leaf.

Figure 2 - Detached peanut leaf packed in a Petri dish.

4.4.1. Preparation of bacterial paste for processing

A colony of *A. rhizogenes* strain K599, previously stored in a freezer at -80C° and transformed with the binary vector pPZP (empty vector) or containing one of the three candidate genes *(pPZP-AdEXLB8, pPZP-AsDUF* or *pPZP-AsLIP),* was grown in a Petri dish containing semi-solid LB medium containing streptomycin (100 mg/mL) and kanamycin (100 mg/mL) at 28°C for 48 hours. An isolated colony was then grown again in semi-solid LB medium with the same antibiotics for a further 48 hours. This culture was resuspended in liquid LB medium containing 50% (v/v) glycerol and spread on Petri dishes containing semi-solid LB medium with the same antibiotics. After growing at 28 °C for 16 hours, all the bacterial culture grown on the plate (called "paste" here) was then collected using a Drigalski loop (Figure 3) and used to inoculate the petiole of peanut leaves or the hypocotyl of soya beans.

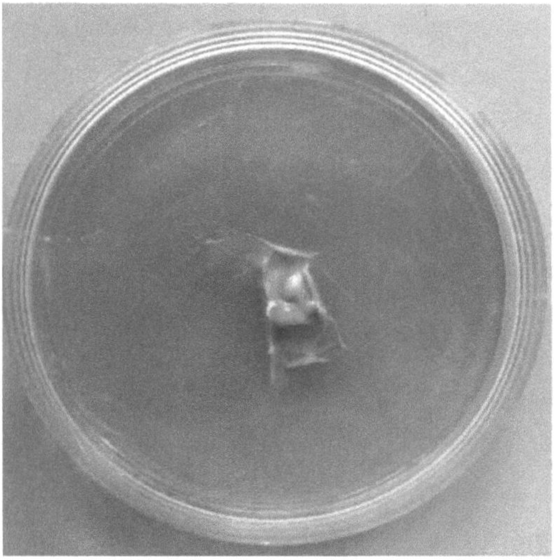

Figure 3 - Obtaining a semi-solid culture ("paste") from a colony of A *rhizogenes* strain K599 containing the binary vector pPZP.

4.4.2. Obtaining transgenic peanut roots

After preparing the plates, the detached peanut leaves were inoculated with the wild strain of *A. rhizogenes* K599 containing the four different binary vectors (pPZP; *pPZP-AdEXLB8, pPZP-AsDUF* or *pPZP-AsLIP) by* making three wounds in the petiole region using a previously sterilised pin soaked in the respective bacterial "paste". Twenty leaves were inoculated for each construct. The leaves were then reconditioned without high humidity in Petri dishes and the wound area covered with cotton wool moistened with distilled water. The plates were kept in a growth chamber at 25 ± 2 °C with a 16-hour photoperiod for 20 days until the transgenic roots grew (hairy roots). The roots were then assessed by observing the fluorescence of the eGFP protein at a wavelength of 488 nm using a Leica M205 FA stereoscopic fluorescence microscope (Leica Microsystem, Germany). Detached leaves with roots expressing little or no fluorescence (considered non-transgenic) were discarded. Leaves with all roots expressing eGFP fluorescence were selected for subsequent inoculation with the nematode. When the same leaf had positive and negative roots, those that did not show fluorescence (negative, non-transgenic) were excised and those that showed fluorescence (positive, transgenic) were kept. The petiole and roots of the selected detached leaves were covered with medium-grained vermiculite (Figure 4) (5 to 8 mm), previously sterilised by autoclaving, and the detached leaves were kept in a growth chamber at 25 ± 2 °C with a 16-hour photoperiod for the duration of the bioassay.

4.6. Inoculation of peanut roots with *Meloidoyne arenaria*

Each detached leaf was inoculated with *M. arenaria* using a 5 mL syringe (Figure 4) containing 1 mL of a suspension of 1,000 J2 individuals. The inoculum was spread evenly over the vermiculite-covered area (petiole and roots; Figure 4). After inoculation, the plates were kept in a growth chamber at 25 ± 2°C with a 16-hour photoperiod.

60 days after inoculation, the roots were carefully removed from the vermiculite, washed in running water and examined under a Leica M205 FA stereoscopic fluorescence microscope (Leica Microsystem, Germany). The transgenic roots (emitting eGFP fluorescence) were collected using a scalpel blade, weighed, stored in Petri dishes containing distilled water and assessed for the presence of galls and egg masses. To do this, the roots were stained with Floxin B (15 mg/L) for 20 minutes to highlight the egg masses according to Taylor and Sasser, 1978. The number of egg masses and galls was then counted using an optical microscope (AxiosKop, Zeiss - Oberkochen, Germany). The data obtained was statistically analysed to compare the means using Student's t-test *($p< 0.05$)* and the comparison between all the data was carried out using ANOVA analysis of variance *($p< 0.05$)*.

Figure 4 - Inoculation of transgenic peanut roots with *M. arenaria.*

4.7. Obtaining composite soya plants

To obtain transgenic soya bean roots *(Glycine max)* by inoculation with the wild strain K599 of *A. rhizogenes*, the cultivar 'Williams 82' was used, a model genotype for studies with *Meloidogyne* spp. and widely used in genetic studies. Sixty soya bean seeds were germinated on germination paper in glass Petri dishes (150x25mm). The plates were kept in a growth chamber at 25 ± 2 °C and a 16-hour photoperiod. After four days of germination, the soya seedlings were inoculated with the wild strain K599 of *A. rhizogenes* containing the four different binary vectors (pPZP; pPZP-AdEXLB8, pPZP-AsDUF or pPZP-AsLIP) by means of wounds made in the region parallel to the vascular bundle of the hypocotyl (Figure 5), with a sterile pin previously soaked in the bacterial "paste". Twenty seedlings were inoculated for each construct.

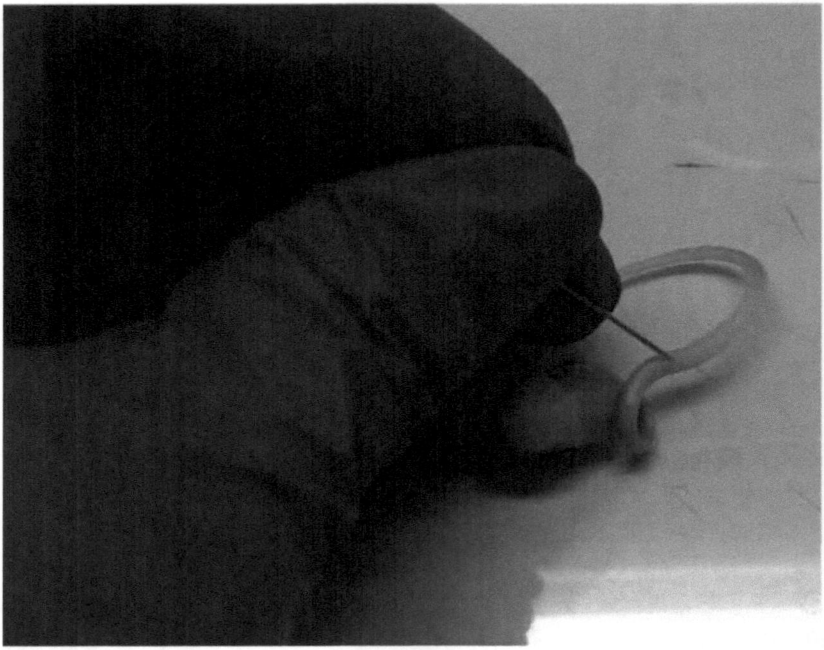

Figure 5 - Inoculation of soya seedlings with A *rhizogenes.*

The seedlings were then placed back in the Petri dishes and covered with filter paper moistened with water. The plates were kept in a growth chamber at 25 ± 2 °C with a 16-hour photoperiod for approximately seven days until the first transformed roots appeared (Figure 6). The roots of the composite plants thus obtained were then selected by observing the fluorescence of the eGFP protein at a wavelength of 488 nm using a stereoscopic fluorescence microscope (Leica Mycrosistem M205 FA). Non-transformed roots were discarded.

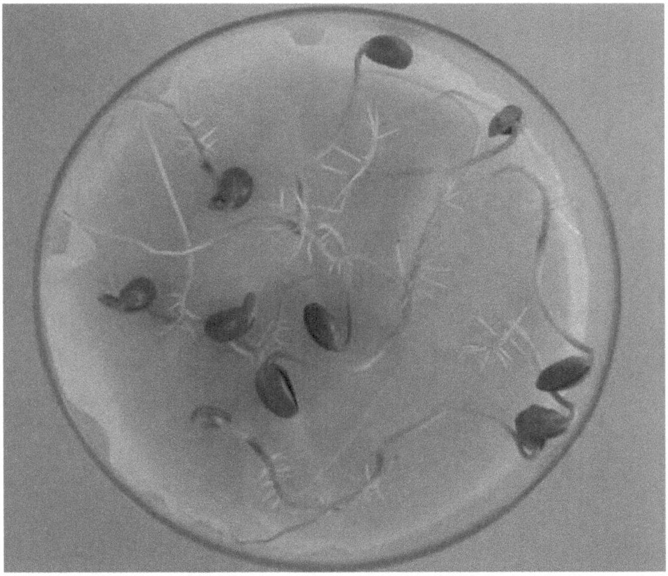

Figure 6 - Transgenic soya roots obtained seven days after transformation with *A. rhizogenes.*

4.8 Inoculation of transgenic soya roots with *Meloidogyne incognita*

4.8.1 Preparing the composite plants for inoculation

After selection, the composite plants (two weeks after transformation with *A. rhizogenes)* with transformed roots (expressing eGFP) were transplanted into 300 mL plastic cups containing the sand-gel mixture and kept in the greenhouse (Figure 7). The gel-sand mixture was previously prepared by dissolving 12 g of Forth planting gel (water-repellent polymer, Tecnutri, Brazil) in 4 L of water and, after a few hours, blended in a blender and mixed with 12 kg of autoclaved fine sand (grain diameter 0.1 mm). The mixture was transferred to the cups after total drying, which lasted approximately one week.

Figure 7 - Composite soya plants five days after transplanting into the gel-sand mixture.

4.8.2 Root inoculation with *M. incognita*

One week after transplanting into the gel and sand mixture, each composite plant was inoculated using a 5 mL syringe containing 1 mL with 1,000 J2 individuals of *M. incognita.* After inoculation, the plants were kept in the greenhouse.

60 days after inoculation, the roots were carefully removed, washed in running water and examined under a Leica M205 FA stereoscopic fluorescence microscope (Leica Microsystem, Germany). The transgenic roots (expressing eGFP) were collected, weighed and stored in Petri dishes to be assessed for the presence of galls and egg masses. Subsequently, the roots were stained with Floxin B (15 mg/L) for 20 minutes to highlight the egg masses, according to Taylore Sasser (1978). The number of egg masses and galls was then counted under an optical microscope (AxiosKop, Zeiss - Oberkochen, Germany). The inoculation bioassay was repeated under the same conditions to analyse a larger number of individuals. The data obtained was statistically

analysed to compare the means using the Student's t-test and all the data was compared using ANOVA.

CHAPTER 5

RESULTS

5.1. Cloning of candidate genes into a binary vector and introduction into the wild strain of *A. rhizogenes*

The ORFs of the *AdEXLB8, AsDUF* and *AsLIP* genes were successfully cloned into the *XhoI* site of the binary vector pPZP, generating the vectors pPZP-AdEXLB8, pPZP-AsDUF and pPZP-AsLIP, respectively (Figure 8). These three vectors and the empty pPZP vector were subsequently introduced into the K599 strain of *A. rhizogenes.* Confirmation of the transformation of the bacteria was carried out using colony PCR (data not shown).

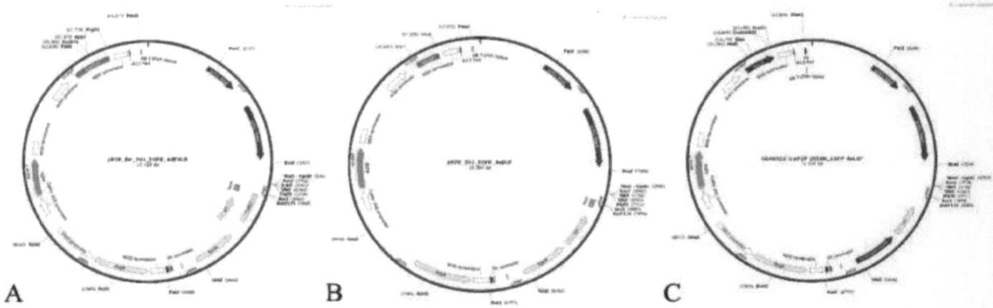

Figure 8 - Map of the vectors used: pZP_201 *RK_AdEXLB8* (A); pZP_201BK_AsZ)tZF (B) and pZP_201BK_As£ZP (C).

5.2. Analysis of transgenic peanut roots inoculated with *M. arenaria*

5.2.1 Transformation with the pPZP vector (control)

Twenty detached peanut leaves were inoculated with the K599 strain of *A. rhizogenes* containing the binary vector pPZP (empty vector). Of these, 18 detached leaves (90%) developed roots with a typical hairy root phenotype 20 days after transformation, with an average of six roots per leaf. Fluorescence microscopy analysis showed that all the roots with this phenotype showed typical fluorescence from the eGFP protein. These results indicate that all the roots developed from inoculating detached peanut leaves with the wild strain of *A. rhizogenes* were transformed with the pPZP binary vector and express the *GFP* transgene. The high transformation efficiency obtained (90%) validates the detached leaf transformation methodology developed in this study. The 18 detached leaves containing transgenic roots were then inoculated with *M. arenaria.*

60 days after inoculation with *M. arenaria,* only the transgenic roots (still expressing eGFP) were evaluated (Figures 9 and 10) by counting galls and egg masses using an optical microscope. To highlight the egg masses, the roots were stained with Floxin B. Of the 18 detached leaves inoculated with *M. arenaria,* 10 (55%) had transgenic roots, i.e. those that emitted fluorescence from the eGFP protein and were therefore selected for analysis of nematode infection. This reduction in the number of detached leaves showing fluorescent roots at 90 days after transformation with *A. rhizogenes* is probably related to the appearance of false negatives during these 70 days, i.e. transgenic roots in which it was not possible to visualise eGFP fluorescence. eGFP fluorescence can be hidden in mature roots due to the reduction in cell density and the thickening of the tissue, which makes it difficult to visualise eGFP in hairy roots. Each of the ten detached leaves contained around six transgenic roots weighing an average of 0.24g (Figure 9 and Appendix 1). Roots transformed with the pPZP vector had an average of 18.9 galls/g root and 13 egg masses/g root (Figures 11 to 13). There was a variation in the total number of galls developed on each transgenic root evaluated (ranging from 9.52 to 26.67 galls/g) or in the egg mass (ranging from 8.00 to 17.86 egg masses/g), indicating that there is an effect of the transgenic event on the development of the disease.

Figure 9 - Transgenic *A. hypogaea* roots obtained 90 days after inoculation with the K599 strain of *A. rhizogenes* (A); transgenic roots under white light (B) and under ultraviolet light (488 nm) (C).

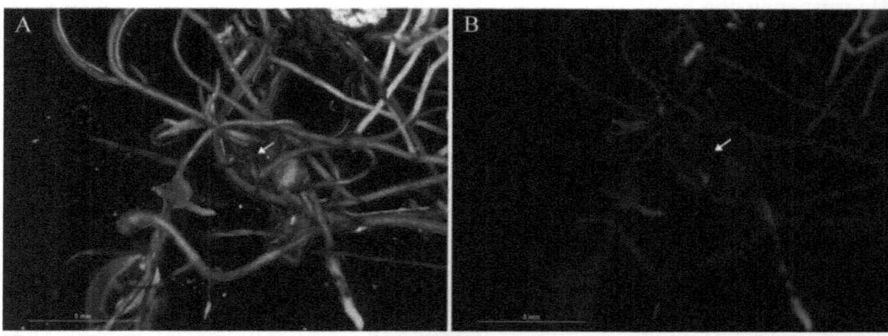

Figure 10 - Transgenic roots expressing eGFP transformed with the empty pPZP vector under white light (A) and ultraviolet light (488 nm) (B). Arrows indicating egg mass.

5.2.2. Transformation with the pPZP-AdEXLB8 vector

Of the twenty detached peanut leaves inoculated with the K599 strain of *A. rhizogenes* containing the pPZP-AdEXLB8 binary vector, 19 (95%) developed roots with a typical hairy root phenotype 20 days after transformation. As with the control with the empty vector, fluorescence microscope analysis showed that all the roots with this phenotype showed typical fluorescence of the eGFP protein. Both the average number of transgenic roots per detached leaf (six roots/leaf) and the transformation efficiency (95 per cent) were very close to those obtained in leaves transformed with the empty vector. These results indicate that the presence of the *EXLB8* gene did not influence the average number of roots per detached leaf and the transformation efficiency. The 19 detached leaves containing transgenic roots were then inoculated with *M. arenaria.*

60 days after inoculation with *M. arenaria,* the roots developed from the detached leaves were once again assessed for the presence of fluorescence. Of the 19 detached leaves inoculated with *M. arenaria,* 14 (73%) had transgenic roots, i.e. that emitted fluorescence from the eGFP protein and were therefore selected for analysis of nematode infection. Each of the 14 detached leaves evaluated contained around six transgenic roots weighing an average of 0.24g, demonstrating that overexpression of the *AdEXLB8* transgene did not influence the average weight and number of roots per detached leaf when compared to roots transformed with the empty vector (Annex 1). Roots transformed with pPZP- AdEXLB8 had an average of 0.26 galls/g and 0.21 egg masses/g (Figure 11). As with the transformation with the empty vector, there was variation between individuals (events), with values ranging from 0 to 3.57 galls/g root and from 0 to 2.54 egg masses/g root (Appendix 1). According to the Student's t-test *(p*< 0.05), roots overexpressing the *AdEXLB8* transgene differed significantly from control roots *(*empty vector) in terms of the number of galls/g (p = 6.74657E-07) and egg masses/g *(p*= 2.7803 5E-07). These results indicate that the *in planta* overexpression of *AdEXLB8* conferred a 97.53% and 98.40% reduction in the number of galls and egg masses, respectively, and a drastic reduction in the development of the infection caused by the nematode compared to the control without *AdEXLB8* (empty vector; Figure 11).

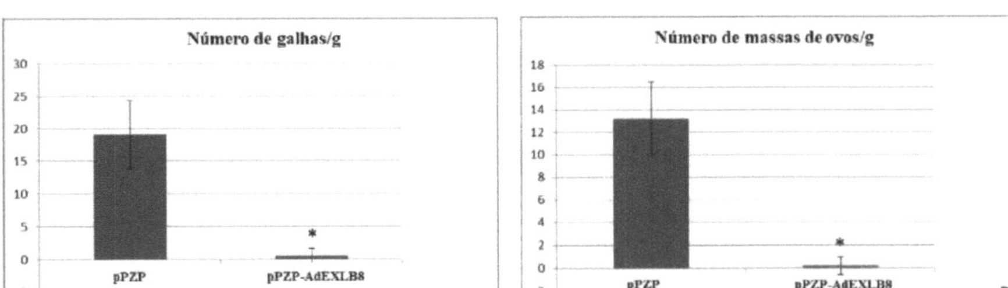

Figure 11 - Average number of galls per gram of root (A) and number of egg masses per gram of root (B) in peanut roots transformed with the empty vector (pPZP) and the pZPZ-AdEXLB8 vector. The * indicates that there was a significant difference between the treatments.

5.2.3 Transformation with the pPZP-AsDUF vector

Of the twenty detached peanut leaves inoculated with the K599 strain of *A. rhizogenes* containing the pPZP-AsDUF binary vector, 16 (80 per cent) developed roots with a typical hairy root phenotype 20 days after transformation. As with the control with the empty vector, fluorescence microscope analysis showed that all the roots with this phenotype showed typical fluorescence of the eGFP protein. Both the average number of transgenic roots per detached leaf (six roots/leaf) and the transformation efficiency (80 per cent) were very close to those obtained in leaves transformed with the empty vector. These results indicate that the presence of the *AsDUF* gene did not influence the average number of roots per detached leaf and the transformation efficiency. The 16 detached leaves containing transgenic roots were then inoculated with *M. arenaria.*

60 days after inoculation with *M. arenaria,* the roots developed from the detached leaves were again assessed for the presence of fluorescence. Of the 16 detached leaves inoculated with *M. arenaria,* 10 (62%) had transgenic roots, i.e. that emitted fluorescence from the eGFP protein and were therefore selected for analysis of nematode infection. Each of the 10 detached leaves evaluated contained around six transgenic roots weighing an average of 0.28g, demonstrating that overexpression of the *AsDUF* transgene did not influence the average weight and number of roots per detached leaf when compared to roots transformed with the empty vector (Annex 1). Roots transformed with pPZP-AsDUF had an average of 2.44 galls/g and 2.15 egg masses/g (Figure 12). As with the transformation with the empty vector, there was variation between individuals (events), with values ranging from 0 to 5.71 galls/g root and from 0 to 5.00 egg masses/g root (appendix 1). According to the Student's t-test ($p < 0.05$), roots overexpressing the *AsDUF* transgene differed significantly from control roots in terms of the number of galls/g ($p = 9.11463E-07$) and egg masses/g ($p =$

1.96593E-07). There was no significant difference between the roots transformed with the pPZP-AsDUF vector and the pPZP-AdEXLB vector by analysis of variance (ANOVA). These results indicate that the *in planta* overexpression of *AsDUF* conferred an 87.17% and 83.67% reduction in the number of galls and egg masses, respectively, and a drastic reduction in the development of the infection caused by the nematode compared to the control without *AsDUF* (Figure 12).

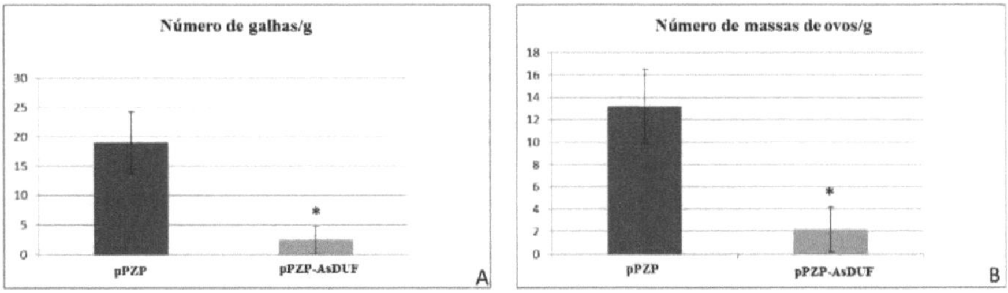

Figure 12 - Average number of galls per gram of root (A) and number of egg masses per gram of root (B) in peanut roots transformed with the empty vector (pPZP) and the pZPZ-AsDUF vector. The * indicates that there was a significant difference between the treatments.

5.2.4Transformation with the pPZP-AsLIP vector

Of the twenty detached peanut leaves inoculated with the K599 strain of *A. rhizogenes* containing the pPZP-AsLIP binary vector, 17 (85%) developed roots with a typical hairy root phenotype 20 days after transformation. As with the control with the empty vector, fluorescence microscope analysis showed that all the roots with this phenotype showed typical fluorescence of the eGFP protein. Both the average number of transgenic roots per detached leaf (six roots/leaf) and the transformation efficiency (85%) were very close to those obtained in leaves transformed with the empty vector. These results indicate that the presence of the *AsLIP* gene did not influence the average number of roots per detached leaf and the transformation efficiency. The 17 detached leaves containing transgenic roots were then inoculated with *M. arenaria.*

After 60 days of inoculation with *M. arenaria,* the roots developed from the detached leaves were again assessed for the presence of fluorescence. Of the 17 detached leaves inoculated with *M. arenaria,* 12 (70%) had transgenic roots and were therefore selected for analysis of nematode infection. Each of the 12 detached leaves evaluated contained around six transgenic roots weighing an average of 0.27g, demonstrating that overexpression of the *AsLIP* transgene did not influence the average weight and number of roots per leaf

when compared to the roots transformed with the empty vector (Appendix 1). Roots transformed with pPZP-AsLIP showed an average of 8.04 galls/g and 6.17 (Figure 13). As with the transformation with the empty vector, there was great variation between individuals (events), with values ranging from 4.00 to 10.00 galls/g root and from 3.70 to 9.38 egg masses/g root (appendix 1). According to the Student's t-test *($p< 0.05$)*, roots overexpressing the *AsLIP* transgene differed significantly from control roots (empty vector) in terms of the number of galls/g (p = 6.04221E-05) and egg masses/g *(p* = 4.98708E-05). A significant difference was observed between the roots transformed with the pPZP-AsLIP vector compared to the roots transformed with the pPZP-AdEXLB and pPZP-AsDUF vectors by analysis of variance (ANOVA). These results indicate that the *in planta* overexpression of *AsLIP* conferred a 57.76% and 46.84% reduction in the number of galls and egg masses, respectively, and a decrease in the development of the infection caused by the nematode compared to the control *without AsLIP* (Figure 13).

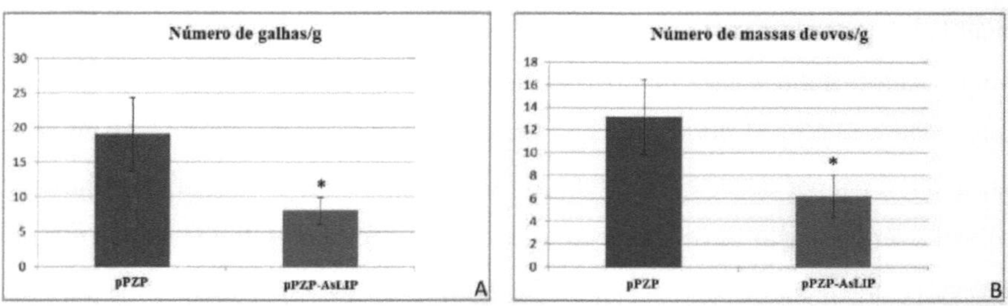

Figure 13 - Average number of galls per gram of root (A) and number of egg masses per gram of root (B) in peanut roots transformed with the empty vector (pPZP) and the pZPZ-AsLIP vector. The * indicates that there was a significant difference between the treatments.

As observed, the transgenic roots overexpressing the candidate genes differed significantly from the control roots in reducing nematode infection (Figure 14).

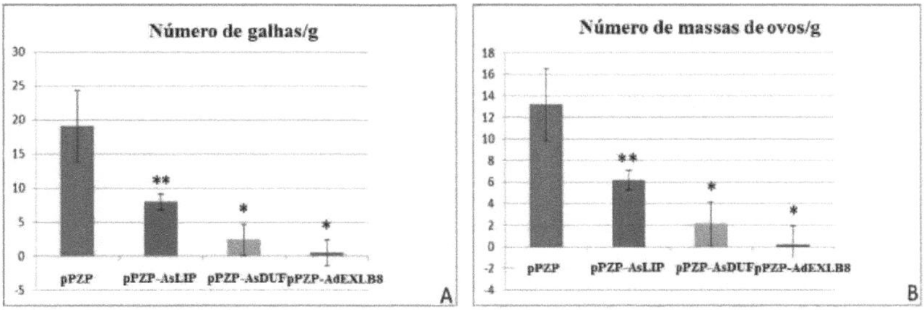

Figure 14 - Average number of galls per gram of root (A) and average number of egg masses per gram of root (B) in all the transformed roots evaluated. The * indicates that there was a significant difference between the treatments.

5.3 Analysis of transgenic soya roots inoculated with *M. incognita*

5.3.1 Transformation with the pPZP vector (control)

The data is based on the total number of plants from the two bioassays carried out. Forty soya seedlings were inoculated with the K599 strain of *A. rhizogenes* containing the binary vector pPZP (empty vector). Of these, 28 plants (70%) developed roots with a typical hair root phenotype seven days after transformation. Fluorescence microscopy analysis showed that 60 per cent of the roots with this phenotype showed fluorescence typical of the eGFP protein. The 28 plants containing transgenic roots were then inoculated with *M. incognita.*

60 days after inoculation with *M. incognita,* only the transgenic roots (still expressing eGFP) were evaluated (Figure 15) by counting galls and egg masses using an optical microscope. To highlight the egg masses, the roots were stained with Floxin B. Of the 28 inoculated plants, 19 (67%) had transgenic roots and were therefore selected for nematode infection analysis. Each of the nineteen plants contained around 60% transgenic roots weighing an average of 2.98g (Annex 2). Roots transformed with the

pPZP showed an average of 16.09 galls/g and 15.42 egg masses/g (Figures 16 to 18). There was variation in the total number of galls developed on each transgenic root evaluated (ranging from 10.32 to 34.62 galls/g root), and in egg masses (ranging from 10.03 to 31.25 egg masses/g root).

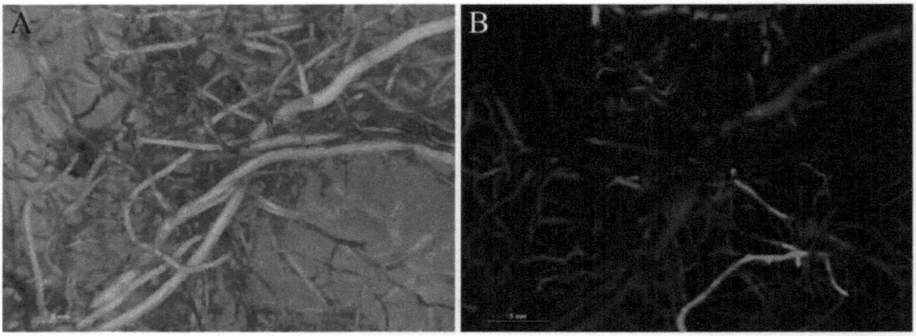

Figure 15 - Transgenic roots of *G.max* obtained 90 days after inoculation with the K599 strain of *A. rhizogenes'*, transgenic roots under white light (A) and under ultraviolet light (488 nm) (B).

5.3.2 Transformation with the pPZP-AdEXLB8 vector

Of the forty soya bean seedlings inoculated with the K599 strain of *A. rhizogenes* containing the pPZP-AdEXLB8 binary vector, 30 (75%) developed roots with a typical hairy root phenotype seven days after transformation. As with the control with the empty vector, fluorescence microscope analysis showed that approximately 60 per cent of the roots with this phenotype showed fluorescence typical of the eGFP protein. Both the average number of transgenic roots per plant (60%) and the transformation efficiency (75%) were very close to those obtained in plants transformed with the empty vector. The 30 plants containing transgenic roots were then inoculated with *M. incognita.*

After 60 days of inoculation, the roots were again assessed for the presence of fluorescence. Of the 30 inoculated plants, 21 (70%) had transgenic roots, i.e. those that emitted fluorescence from the eGFP protein and were therefore selected for analysis of nematode infection. Each of the 21 plants evaluated contained around 60% transgenic roots weighing an average of 3.60g, demonstrating that overexpression of the *AdEXLB8* transgene did not influence the average weight and number of roots per plant when compared to roots transformed with the empty vector (Annex 2). Roots transformed with pPZP-AdEXLB8 had an average of 0.68 galls/g and 0.66 egg masses/g (Figure 16). As with the transformation with the empty vector, there was variation between individuals, with values ranging from 0 to 1.34 galls/g root and from

0 to 1.79 egg masses/g root (Annex 2).According to the Student's t-test ($p< 0.05$), roots overexpressing the *AdEXLB8* transgene differed significantly from control roots in terms of the number of galls/g (p = 1.79536E-09) and egg masses/g (p= 2.17839E-10). These results indicate that the *in planta* overexpression of *AdEXLB8* conferred a 95.77% and 95.90% reduction in the number of galls and egg masses, respectively, and a drastic reduction in the development of the infection caused by the nematode compared to the control without *AdEXLB8* (Figure 19 A and C).

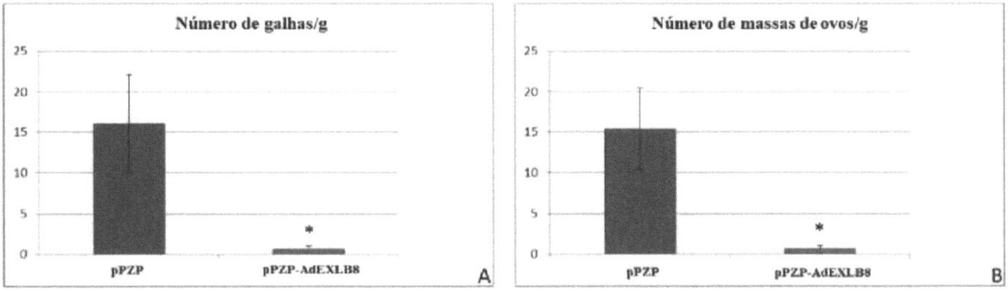

Figure 16 - Average number of galls per gram of root (A) and number of egg masses per gram of root (B) in soya roots transformed with the empty vector (pPZP) and the pZPZ-AdEXLB8 vector. The * indicates that there was a significant difference between the treatments.

5.3.3 Transformation with the pPZP-AsDUF vector

Of the forty soya seedlings inoculated with *A. rhizogenes* strain K599 containing the binary vector pPZP-AsDUF, 25 (62%) developed roots

with a typical hairy root phenotype, seven days after transformation. As with the control with the empty vector, fluorescence microscope analysis showed that approximately 60 per cent of the roots with this phenotype showed fluorescence typical of the eGFP protein. Both the average number of transgenic roots per plant (60%) and the transformation efficiency (62%) were close to those obtained in plants transformed with the empty vector. The 25 plants containing transgenic roots were then inoculated with *M. incognita.*

After 60 days of inoculation, the roots developed from the detached leaves were again assessed for the presence of fluorescence. Of the 25 plants inoculated with *M. incognita,* 15 (60%) had transgenic roots, i.e. those that emitted fluorescence from the eGFP protein and were therefore selected for analysis of nematode infection. Each of the 15 plants evaluated contained around 60% of transgenic roots weighing an average of 3.63g, showing that overexpression of the *AsDUF* transgene did not influence the average weight and number of roots per plant when compared to roots transformed with the empty vector (Annex 2). Roots transformed with pPZP-AsDUF had an average of 2.39 galls/g and 2.52 egg masses/g (Figure 17). As with the transformation with the empty vector, there was great variation between individuals (events), with values ranging from 0 to 3.52 galls/g root and from 0 to 3.49 egg masses/g root (appendix 2). According to the Student's t-test *($p< 0.05$)*, roots overexpressing the *AsDUF* transgene differed significantly from control roots in terms of the number of galls/g (p = 8.06661E-09) and egg masses/g (p= 1.34137E-09). These results indicate that the *in planta* overexpression of *AsDUF resulted in an* 85.15% and 84.29% reduction in the number of galls and egg masses, respectively, and a drastic reduction in the development of the infection caused by the nematode compared to the control without *AsDUF* (Figures 17 and 19 A and B). There was no significant difference between the roots transformed with the pPZP-AsDUF vector and the pPZP-AdEXLB vector by analysis of variance (ANOVA).

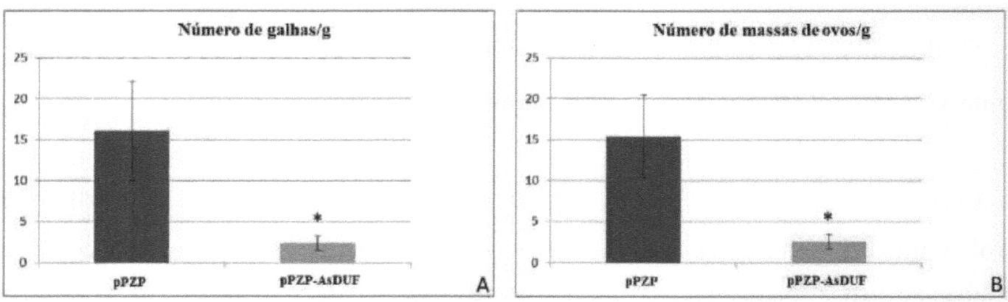

Figura 17 - Average number of galls per gram of root (A) and number of egg masses per gram of root (B) in soya roots transformed with the empty vector (pPZP) and the pZPZ-AsDUF vector. The * indicates that there was a significant difference between the treatments.

As observed, the transgenic roots overexpressing the candidate genes differed significantly from the control roots in reducing nematode infection (Figures 18 and 19).

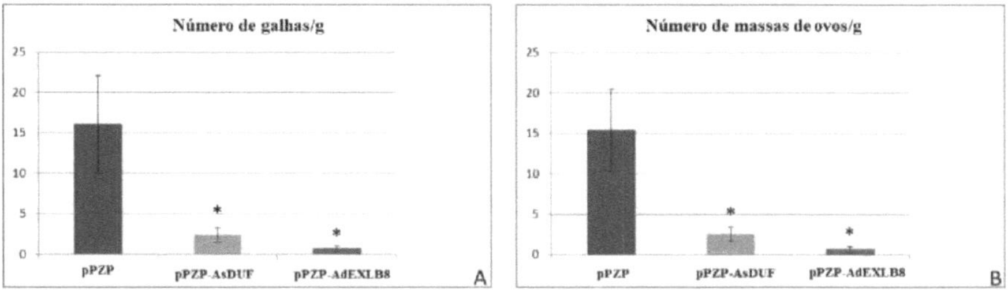

Figura 18 - Average number of galls per gram of root (A) and average number of egg masses per gram of root (B) in all the transformed roots evaluated.

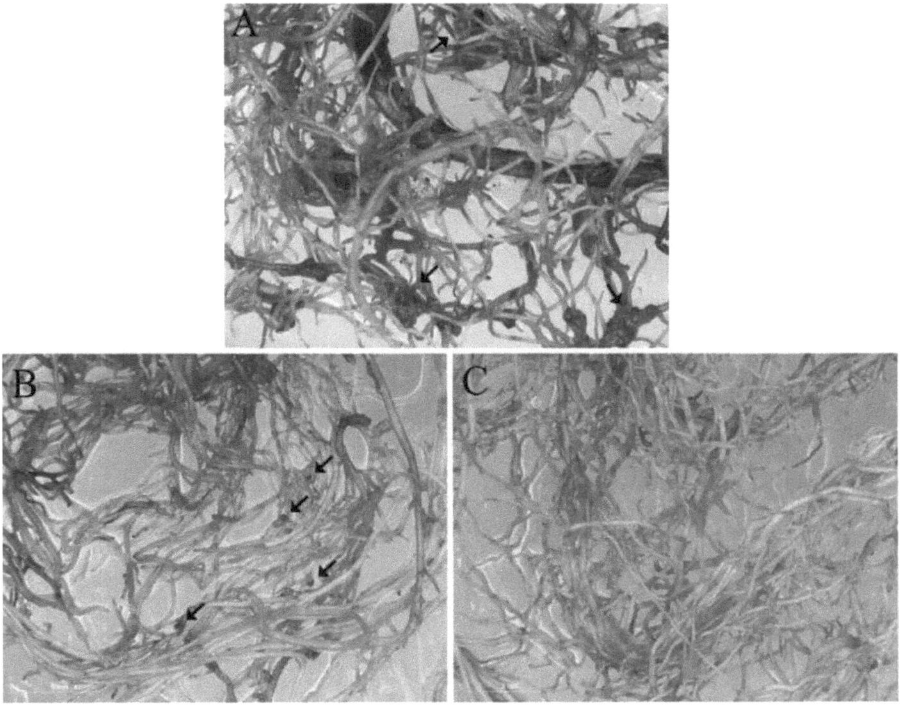

Figura 19 - Transgenic soya roots after 60 days of inoculation with *M.incognita.* Roots transformed with the pPZP vector (control) (A); roots transformed with pPZP-AsDUF (B) and with pPZP-AdEXLB8 (C). Arrows indicating egg masses.

CHAPTER 6

DISCUSSION

6.1. Induction of transgenic roots in detached peanut leaves

The efficiency of peanut transformation with the wild strain K599 of *A. rhizogenes* obtained here (90%) by inoculating petioles on detached leaves was higher than the average efficiency previously observed for obtaining peanut composite plants (Akasaka et al., 1998; Geng et al., 2012; Chu et al., 2014). In order to obtain a transformation with *A. rhizogenes with* high efficiency (above 80%), some parameters must be considered, such as: type of explant; density and growth stage of the bacterial cells; strain of *A. rhizogenes* and humidity. Several types of peanut explants have already been tested for *A. rhizogenes* inoculation, such as hypocotyl, embryonic axis, epicotyl and leaves, and all of them showed a transformation efficiency of less than 75% (Akasaka et al., 1998; Chu et al., 2014; Liu et al., 2016). The only exception was the petiole in detached leaves with 91% transformation efficiency (Liu et al., 2016) which is similar to the efficiency found in the present work. The lowest efficiency (14%) was obtained using hypocotyls under *in vitro* conditions.

The first study to transform peanuts with wild strains of *A. rhizogenes in* order to obtain composite plants was carried out by Akasaka and collaborators (1998), where the morphology of the hair-like roots and the formation of root nodules were observed. In this study, the transformation efficiency of epicotyls was 62 per cent. More recent work (Geng et al., 2012) showed similar efficiency when using embryonic axes to assess the insecticidal activity of a synthetic gene *{crySEal)* possibly involved in resistance to *Holotrichia parallela* in transgenic peanut roots. In 2014, transformation efficiency was increased to 75% using peanut hypocotyls (Chu et al., 2014). In addition to the biotic stresses mentioned above, peanut composite plants have also been used recently to study the responses of transgenic roots to abiotic stress, in particular drought (Liu et al., 2016). Composite plants were obtained with high transformation efficiency

(91%) overexpressing the following genes: *AhAREBl,* a transcription factor isolated from peanut and related to drought resistance, and four stress-responsive genes (the transcription factors *WRKY33 and MYB92,* an abscisic acid receptor *PYL5* and a dehydrin 2 *DHN2. In vitro* cultures of transgenic peanut roots have also been exploited for the production and secretion of high levels of resveratrol, a metabolite with antioxidant and anticancer properties (Medina-Bolivar et al., 2007).

One limitation to the successful induction of transgenic roots in peanut is the cell density and developmental stage of the *Agrobacterium* strain (Geng et al., 2012; Bansal et al., 2014). The bacterial paste used in the protocol developed here combined a high density of bacterial cells with a recent *Agrobacterium* culture (in exponential growth phase), allowing for efficient transformation (Kereszt et al., 2007; Chu et al., 2014). The paste used in the transformation maintained exponential growth of *Agrobacterium* even after inoculation into the petiole. Sonia et al. (2007) observed adverse effects of accentuated bacterial growth, such as the production of toxic compounds that cause necrosis in the explant, reducing the transformation rate. The *A. rhizogenes* paste is introduced into the vascular tissue inside the petiole by means of a needle, thus bringing the bacterium into contact with the cells of the vascular tissue that have the competence to differentiate into roots (cambium and vascular parenchyma) (Chen et al., 2016). In addition, this inoculation process increases the production of phenolic compounds in the wound area, which attracts the bacteria by chemotaxis to the cells competent for differentiation, increasing transformation efficiency (Chandra, 2012).

In addition to the K599 strain of *A. rhizogenes* used here in the transformation, many other pathogenic wild strains (ATCC 15834, A4 and MAFF-02- 10266) and engineered strains (RI000, RI 601 and RI200) of *A. rhizogenes* have also been used to induce hairy roots in peanuts (Akasaka et al., 1998; Kim et al., 2010 Chu et al., 2014). However, only the wild strain K599 has been successfully used in studies of peanut-gall nematode interaction (Chu et al., 2014), attesting that nematodes can complete their life cycle in transgenic roots induced by this strain. One of the main restrictions in the use of hairy roots for gene functional analysis studies is their morphological variability (plagiotropic and rapid growth, high degree of lateral branching and profusion of root hairs) due to the hormonal imbalance in the root promoted by the expression of T-DNA oncongenes (Mehrotra et al., 2015). This phenotype, known as *hairy root,* is strongly associated with the *A. rhizogenes* strain used and can make it difficult to analyse the pleiotropic effects induced by transgene expression in transformed roots. Despite their phenotype, hairy roots generally behave like differentiated and functional root organs and can be exploited for root biology studies (Plovie et al., 2003), as in the case of the K599 strain used in this study.

Finally, the chamber set up to maintain high humidity inside the Petri dish seemed to improve transformation efficiency, since high humidity is necessary in the days following inoculation for the correct formation and development of transgenic roots (Estrada-Navarrete et al., 2007; Kereszt et al., 2007; Chu et al., 2014). The use of a high humidity chamber in a Petri dish is only possible when small explants are used, as in the case of detached leaves, making it unfeasible for whole plants, as in the case of composite plants (Akasaka et al., 1998; Chu et al., 2014). Furthermore, the use of Petri dishes, each containing up to two detached leaves, considerably reduces the space used for the assay, allowing large-scale analysis, since a single leaf can produce an average of six transgenic roots, each representing an independent transformation event. In addition, being an *ex vitro* methodology, the methodology developed here does not require tissue culture infrastructure and material, greatly reducing the costs of its realisation.

6.1.1. Inoculation of *M. arenaria* in transgenic peanut roots

Transgenic roots obtained after inoculation of the petiole of detached peanut leaves were tested for infection with the gall-forming nematode *M. arenaria.* The *AdEXLB8, AsDUF and AsLIP* genes, isolated from wild peanut species resistant to *M. arenaria (A. duranensis* and *A. stenosperma),* were chosen as candidates to be overexpressed in transgenic roots of the cultivated species *(A. hypogaea).* The transgenic peanut roots transformed with the empty vector were successfully infected by *M. arenaria,* observing the symptoms of infection, such as the establishment of the feeding site and the production of egg masses at the end of its life cycle, 60 days after inoculation. These results suggest that the transgenic roots developed in the petiole of detached peanut leaves after inoculation with *A. rhizogenes* constitute a good model for the *in planta* validation of candidate genes and can be successfully used for nematode infection bioassays.

Transgenic peanut roots overexpressing the three candidate genes *(AdEXLB8, AsDUF and AsLIP)* showed a significant reduction in the number of galls and egg masses at 60 days after inoculation. These genes can thus confer, via transgenesis, resistance to infection by *M. arenaria* in the peanut crop, and corroborated previous studies showing that overexpression of *AdEXLB8* in roots of composite soya plants significantly reduced infection by *M. javanica* (Guimaraes et al, 2017). Over the last few years, transgenic roots have been widely exploited for studies related to nematode infection in various crops such as peanuts, soya, coffee, melons and beans (Alpizar et al., 2006; Estrada-Navarrete et al., 2007; Pak et al., 2009; Bosselut et al., 2011; Matthews et al., 2013). Due to their simplicity, efficiency and low cost, these roots have become the method of choice for large-scale analysis of candidate genes related to interactions between plants and root microorganisms (pathogenic or symbiotic), such as nematodes, fungi and bacteria. In general, the process of producing a transgenic plant is long, labour-intensive and expensive, especially for plants that are recalcitrant to genetic transformation, such as peanuts. In addition, the *ex vitro* induction of transgenic roots, as described here, is more suitable for studies of plant-nematode interactions, as it avoids the critical steps of sterilising nematodes and maintaining a complex axenic system made up of three organisms: plant, nematode and bacteria (Narayanan et al., 1999).

In summary, the main advantages of using hairy roots obtained by the transformation methodology of petioles from detached peanut leaves developed here to validate candidate genes *in planta* are:

(i) Space saving, since all the steps are carried out in Petri dishes and don't require whole plants;

(ii) Saving time, since the hairy roots can be used for phenotyping 20 days after inoculation with *A. rhizogenes',*

(iii) *Ex vitro* conditions, which allow the use of non-sterile microorganisms (pathogens or symbionts) for root inoculation;

(iv) Low cost, as no tissue culture facilities are needed;

(v) Large-scale analysis, due to the fact that a single detached leaf can produce an average of six transgenic roots and each root represents an independent transformation event;

(vi) High transformation efficiency.

6.2. Induction of transgenic roots in composite soya plants

Soya transformation efficiency was assessed according to the number of plants that showed at least one transformed root (observation of eGFP protein expression under a microscope), in relation to the number of transformed plants. The average efficiency obtained here was 69%, similar to the work by Kuma et al. (2015), who used the same soya genotype ('Williams 82') in which the average transformation efficiency was 55%. This efficiency is higher than previous studies, in which the average efficiency did not exceed 34%, using the same agroinfiltration methodology described here, but with other genotypes, perhaps more recalcitrant to transformation than 'Willians 82', such as 'Jack'; 'KS4704' and 'Bragg' (Li et al., 2010; Lin et al., 2011). Ninety days after transformation with *A. rhizogenes,* only 66.2 per cent of the plants (55 out of 83 plants) that initially contained transformed roots still had eGFP-positive roots. This result is similar to that observed by Alzohairy et al. (2013), who found a yield of 65 to 70 per cent of transformed plants after approximately 40 days of transformation. Collier et al. (2005) observed that among the transformed soya plants (with at least one transformed root), only 50% of the roots were eGFP positive. This result corroborates that observed in the present study, where 60 per cent of the roots of the plants that showed the hairy root phenotype showed typical fluorescence of the eGFP protein. The time taken to obtain transformed roots ready for inoculation with *M. incognita* observed here (25 days after germination) is shorter than that observed in previous studies (Ibrahim et al., 2011; Alzohairy et al., 2013), which required approximately 40 and 63 days, respectively, to obtain plants with transgenic roots ready to be inoculated with nematodes.

As with detached peanut leaves, the way in which the bacterial paste is inoculated, parallel to the vascular bundle of the hypocotyl of soya seedlings, can considerably influence transformation efficiency. Inoculating the bacteria superficially into the plant tissue can result in low transformation efficiency (Li et al., 2010; Lin et al., 2011). Li et al. (2010) used the method of transforming soya seedlings by infiltration of *A.rhizogenes,* using hygromycin *in the in vitro* selection stages of the transformed roots for a period of 2 to 4 weeks, *in* order to eliminate and/or inhibit the appearance of non-transformed roots and select only the transgenic ones. In the present study, there was no selection by the antibiotic hygromycin, as the transgenic events were selected by the emission of eGFP fluorescence, and all the steps of the protocol were carried out under *ex vitro* conditions, making the work less laborious. Kuma et al. (2015) showed that the use of hygromycin as a selective agent was not able to prevent the formation and/or eliminate non-transgenic roots efficiently after a period of up to 10 days of exposure to the antibiotic. In the present methodology, the detection and elimination of escapes was carried out by excising non-transgenic roots by observing the fluorescence of the eGFP protein. Transformed roots were easily identified by the fluorescence emitted under a microscope, while non-transformed roots showed no signal. The use of reporter genes, such as eGFP, for *in vivo* screening of transgenic events is widely described in the literature (Lacorte and Barros, 2015) and in particular to identify transgenic roots in composite soya plants (Collier et al., 2005; Li et al. 2010; Ibrahim et al. 2011).

6.2.1. Inoculation of A. *incognita in* composite soya plants

The transgenic roots of composite soya plants were successfully infected by the gall-forming nematode *M. incognita,* with establishment of the feeding site and production of egg masses at the end of its life cycle. Kuma et al. (2015) also successfully infected transgenic soya roots with nematodes of the species *M. javanica,* at juvenile development stage 2, and confirmed infection in roots collected eight days after inoculation. These results, together with those presented here, suggest that hairy roots developed in composite soya plants constitute a good model for the *in planta* validation of candidate genes and can be successfully used for nematode infection bioassays.Transgenic soya roots overexpressing the candidate genes *AdEXLB8* and *AsDUF* showed a significant reduction in the number of galls and egg masses of *M. incognita,* as observed in transgenic peanut roots overexpressing the same genes after infection with *M. arenaria.* These results indicate that overexpression of the *AdEXLB8* and *AsDUF* genes may confer resistance to infection by these two species and, more broadly, to gall-forming nematodes *(Meloidogyne* spp.). This hypothesis is corroborated by previous studies by our team which showed that overexpression of *AdEXLB8* in roots of composite soya plants also significantly reduced infection by a third species of *Meloidogyne, M. javanica* (Guimaraes et al., 2017).

6.3. Candidate genes

Over the last few years, hairy roots obtained by infection with wild strains of *A. rhizogenes* have been widely exploited for studies of plant-nematode interaction in many crops, including the *in vivo* validation of candidate genes for resistance to the root-knot nematode *Meloidogyne* spp. (Alpizar et al., 2006; Bosselut et al., 2011; Chu et al., 2014; Guimaraes et al., 2017). The candidate genes validated in this study in peanut and soya showed similar rates of reduction in infection caused by the nematodes *M. arenaria* and *M. incognita,* respectively. Overexpression of the *AdEXLB8* gene conferred a large reduction in the number of galls and egg masses compared to the control. In addition, this gene showed a greater reduction in the symptoms evaluated when compared to the other two candidate genes *(AsDUF* and *AsLIP).* The *AdEXLB8* gene encodes an Expansin-like B isolated *from A. duranensis,* in a previous report the expression of *EXLB8* was associated with nematode resistance (Guimaraes et al., 2017). Expansins are proteins involved in cell wall plasticity in response to environmental stimuli such as attack by phytopathogens (McQueen-Mason et al., 1992; Marowa et al., 2016) and an increase in their expression has been observed in roots exposed to different stresses (Guimaraes et al., 2015; Brasileiro et al., 2015). The second gene studied here, *AsDUF,* encodes a protein that contains a conserved DUF domain *(Domain of Unknown Functiori)* belonging to a family of unknown function (Gholizadeh et al., 2011).Plants, in response to different stresses, activate various proteins that are probably involved in their defence mechanisms. These proteins are found in response to various types of biotic and abiotic stresses (Gholizadeh et al., 2011). Overexpression of *AsDUF* caused a significant reduction in the symptoms caused by the gall-forming nematodes *M. arenaria* and *M. incognita in* transgenic roots, with a very similar decrease in the number of galls and egg masses in peanut and soya when compared to the control. The third candidate gene studied here, *AsLIP,* was tested only in transgenic peanut roots where its overexpression significantly reduced the number of galls and egg masses. However, despite inducing a reduction in disease symptoms, the *AsLIP* gene differed significantly from the other candidate genes *(AdEXLB8* and *AsDUF)* and was therefore not selected for further evaluation in transgenic soya roots. High levels of salicylic acid, which often lead to RH, are regulated by proteins such as Lipocalins, which are involved in modulating tolerance to oxidative stress (Zinovieva et al., 2011).Plant Lipocalins have been shown to play an important role in removing potentially harmful molecules, ROS, known to be induced by stresses related to extreme temperatures and excess light (Charron et al.,2008).

The similarity of the results obtained in peanut and soya suggests that the overexpression of the candidate genes *AdEXLB8 and AsDUF* in transgenic roots of these legumes maintains a pattern in the response to infection and the development of the disease caused by the gall-forming nematodes, *M. arenaria* and *M. incognita.* The discovery of new genes involved in plant-nematode interaction, and in particular those that may be involved in resistance to gall nematodes, is of great interest to peanut and soybean breeding programmes. There are currently peanut cultivars resistant to *M. arenaria,* such as COAN, Nematan and Tifguard, which were developed by introgressing a chromosomal region from the wild peanut species *A. cardenasii* (Holbrook et al., 2003; Tillman., 2010). *In* relation to soya, there are already cultivars resistant to *M. incognita* such as BRS282, BRS256RR, BRS260, developed by Embrapa Soja (Embrapa, 2008). However, nematode species can develop the ability to break the resistance of cultivars, making it necessary to continue searching for new sources of resistance to the *Meloidogyne* populations that cause significant damage to these crops. Therefore, it is desirable to identify new genes related to defence in wild *Arachis* species (which may also be useful for other legumes), in order to expand the sources of resistance implanted in peanuts and also guarantee greater durability of this resistance.

CHAPTER 7

CONCLUSIONS

- An unprecedented, simple, efficient, fast, low-cost and space-saving method has been established for obtaining transgenic roots from alternative explants (detached leaves) of peanuts;

- An efficient methodology for inoculating gall-forming nematodes *(Meloidogyne* spp.) was established in hairy roots developed from detached peanut leaves and in composite soya plants;

- Peanut and soya bean hairy roots are a good model for the *in planta* validation of candidate genes and can be successfully used for plant-nematode interaction studies;

- Overexpression of the *AdEXLB8, AsDUF* and *AsLIP* genes in transgenic roots was able to significantly reduce the infection caused by two species of gall-forming nematodes *(M. arenaria* and *M. incognita)* in different legume species (peanut and soya).

CHAPTER 8

BIBLIOGRAPHICAL REFERENCES

Akasaka, Y., Mii, M., Daimon, H., (1998). Morphological Alterations and Root Nodule Formation inAgrobacterium rhizogenes-mediated Transgenic Hairy Roots of Peanut *(Arachis hypogaea* L.). *Ann Bot,* 81:355-362.

Alpizar, E., Dechamp, E., Espeout, S., Royer, M., Lecouls, A. C., Nicole, M., Bertrand, B., Lashermes, P., Etienne, H., (2006). Efficient production of *Agrobacterium rhizogenes-transformed root* and composite plants for studying gene expression in coffee root. *Plant Cell,* Rep 25:959-967.

Alzohairy, A. M., Macdonald, M. H., Matthews, B. F. (2013). The pjan25 vector series: an enhancement of the Gateway'compatible vector pGWB533 for broader promoter testing applications. *Plasmid. N.* 69. n. 3. p. 249-56.

Bailey, J. (2002). Peanut Disese Management. North Carolina *Peanut Production Guide.* Chapter 6. 19pp.

Bansal, M., Kumar, A., Sudhakara-Reddy. M. (2014). Influence of *Agrobacterium rhizogenes* strains on hairy root induction and "bacoside A" production from *Bacopa monnieri* (L.) Wettst. *Acta Physiol Plant* 36:2793-2801.

Baxter, A., Mittler, R., Suzuki, N. (2014). ROS as key players in plant stress signalling. *J. Exp. Bot.,* 65,1229-1240.

Bertioli, D. J.,Leal-Bertioli, S. C. M., Carneiro, R. G., Valls, J.F.M., Fávero, A. P., Bruzzi, M., Vello, N. A., Moraes, S. and Guimaraes, P. M. (2000). The Brazilian *Arachis* germplasm bank: its diversity, disease resistances and resistance gene analogues. *Molecular Genetics of Model Legumes,* p. 24-28.

Bevan, M. (1984). Binary Agrobacterium vectors for plant transformation. *Nuc. Acid Res.* 12, 8711-8721.

Boneti, J. I. S., Ferraz, S. (1981). Modification of the Hussey & Barker method for extracting Meloidogyne exigua eggs from coffee roots. *Fitopatologia Brasileira,* v. 6, n. 3, 1981.

Bosselut N., Ghelder, C. V., Claverie, M., Voisin, R., Onesto, J. P., Rosso, M. N., Esmenjaud, D. (2011) *Agrobacterium rhizogenes-medÃaieá* transformation of *Prunus* as an altemative for gene functional analysis in Hairy-root and composite plants. *Plant Cell* Rep 30:1313-1326.

Brasileiro A. C. M., Carneiro V. T. C. (2013).Afawwa/ *of plant genetic transformation,* 2ª Edition edn. Embrapa-SPI/Embrapa-Cenargen, Brasilia, Brazil.

Brasileiro, A.C.M., Morgante, C.V., Araújo, A.C.G., Leal-Bertioli, S.C.M., Silva, A.K., Martins, A.C.Q., Vinson, C.C., Santos, C.M.R., Bonfim, O., Togawa, R.C., Saraiva, M.A.P., Bertioli, D.J., Guimaraes, P.M. (2015). Transcriptome Profiling of

Wild *Arachis* from Water-Limited Environments Uncovers Drought Tolerance Candidate Genes. Plant Mol Biol Rep, 33: 1876.

Brunings, A. M., Datnoff, L. E., Ma, J. F., Mitani, N., Nagamura, Y., Rathinasabapathi, B., Kirst, M. (2009). Differential gene expression of rice in response to Silicon and riceblast *hingusMagnaporthe oryzae. Annals of Applied Biology,* v.155, p. 161-170.

Cai, D., Kleine, M., Kifle, S., Harloff, H.J., Sandal, N.N.,Marcker, K.A., Klein-Lankhorst, R.M., Salentijn, E.M., Lange, W., Stiekema, W.J., Wyss, U., Grundler, F.M., Jung, C. (1997). Positional cloning of a gene for nematode resistance in sugar beet. *Science* 275: 832-834.

Chandra, S., (2012). Natural plant genetic engineer Agrobacterium rhizogenes: role of T- DNA in plant secondary metabolism. 407-415.

Charchar, J.M., Boiteux, L.S., Giordano, L.B. (2004).Epidemics of *Meloidogyne brasilienses* on processing tomato hybrids carrying the *Mi* (root-knot nematode resistance) gene in Central Brazil. *Summa Phytopathologica, N.* 30, p. 108.

Charron, J-B. F., Ouellet, F., Houde, M., Sarhan, F. (2008). The plant apolipoproteinD ortholog protects *Arabidopsis* against oxidative stress. *BMC PlantBiology* 8(1), 86.

Chen, X., Cheng, J., Chen, L., Zhang, G., Huang, H., Zhang, Y., Xu, L. (2016). Auxin-Independent NAC pathway acts in response to explant-specific wounding and promotes root tip emergence during de novo root organogenesis in *Arabidopsis. Plant Physiol,* 170:2136-45.

Chu, Y., Guimaraes, L. A., Wu, C. L., Timper, P., Holbrook, C. C., Ozias-Akins, P. (2014). A technique to study *Meloidogyne arenaria* resistance in *Agrobacterium rhizogenesAxansiorvíieá* peanut. *Plant Dis.* 98:1292-1299.

Collier, R., Fuchs, B.,Walter, N., Kevin, W. L., Taylor, C.G. (2005). Ex vitro composite plants: an inexpensive, rapid method for root biology. *The Plant journal: for cell and molecular biology.* v. 43, n. 3, p. 449-57.

CONAB (2004). Conjuntura Semanal. Special 16-20/09. *Ministry of Agriculture, Livestock and Supply. National Supply Company - CONAB,* 2004.

CONAB (2006). Conjuntura Semanal. Special 24-28/06. *Ministry of Agriculture, Livestock and Supply. National Supply Company - CONAB,* 2006.

Cosgrove, D. J. (2015). Plant expansins: diversity and interactions with plant cell walls. *Current Opinion in Plant Biology, N.* 25 pp. 162-172.

Deng, Y., Mao, G., Stutz, W., Yu, O. (2011). Generation of composite plants in *Medicago truncatula* used for nodulation assays. J. *Visual Exp.* 49:2633.

Eisenback, J. D., Bemard, E. C., Starr, J. L., Lee, T. A., Tomaszewski, E. K. (2003). *Meloidogyne haplanaria* n. sp. (Nematoda: Meloidogynidae), a root-knot nematode parasitising peanut in Texas. *Journal ofNematology* 35:395-403.

Embrapa, (2008). Embrapa agency for technological information (ageitec), Gall nematode *(Meloidogyne* spp.). Accessed January 2017, available at https://www.agencia.cnptia.embrapa.br/gestor/soja/arvore/CONT0Q0gcglrznk02wx5 o kOrofsmq 1 qran 1 w.html

Estrada-Navarrete, G., Alvarado-Affantranger, X., Olivares, J. E., Díaz-Camino, C., Santana O, Murillo, E., Guillén, G., Sánchez-Guevara, N., Acosta, J., Quinto, C., Li, D.,Gresshoff, P. M., Sánchez F. (2006). *Agrobacterium rhizogenes* transformation of the *Phaseolus* spp.: A tool for functional genomics. *Mol. Plant Microbe Interact.* 19 1385-1393.

Estrada-Navarrete, G., Alvarado-Affantranger, X., Olivares, J.E., Guillén, G., Díaz-Camino, C., Campos, F., Quinto, C.,Gresshoff, P. M., Sánchez, F.(2007).Fast, efficient and reproducible genetic transformation of *Phaseolus spp.* by *Agrobacterium rhizogenes. Nat Protoc* 2:1819-1824.

Evangelista, A. R., Lopes, J., Abreu, J. G., Neto, P. C. and Fraga, A. C. (2004). Chemical composition of peanut *(Arachis hypogaea* L.) and castor bean *(Ricinus communis* L.) cake - cold oil extraction. I° Congresso de Plantas Oleaginosas, Óleos Vegetais e Biodiesel. Varginha - MG, Brazil.

Fávero, A. P. (2004). Crossability between wild *Arachis* species for the introgression of disease resistance genes into cultivated peanut. PhD thesis. ESALQ, Piracicaba.

Geng, L., Niu, L., Gresshoff, P.M., Shu, C., Song, F., Huang, D., Zhang, J. (2012). Efficient production of *Agrobacterium rhizogenes-Xxwxsíoxmeà* roots and composite plants in peanut *(Arachis hypogaea L.). Plant Cell, Tissue and Organ Culture (PCTOC)* 109:491-500.

Gheysen, G., Fenoll, C. (2002). Gene expression in nematode feeding sites. *Annual Review of Phytopathology,* 40, 191-219.

Gholizadeh, A. (2011). Heterologous Expression of Stress-Responsive DUF538 Domain Containing Protein and its Morpho-Biochemical Consequences. *The Protein Journal,* v. 30pp 351-358.

Gregory, W. C., Krapovickas, A.,Gregory, M. P. (1980). Structure, variation, evolution, and classification in *Arachis.* In: Summerfield, R.J.; Bunting, A.H. (Ed.) Advances in Legume Science. *Kew, Ryal Botanical Gardens,* p.469-481.

Guimaraes P. M., Brasileiro A. C. M., Proite K., de Araújo A., Leal-Bertioli S., Pic-Taylor A., Silva, F. R., Morgante, C. V., Ribeiro, S. G.,Bertioli, D. J. (2010). A Study of Gene Expression in the Nematode Resistant Wild Peanut Relative, *Arachis stenosperma,* in Response to Challenge with *Meloidogyne arenaria. Tropical Plant Biol.* 3(4): 183-92.

Guimaraes, P. M., Guimaraes, L. A., Morgante, C. V., Silva Jr, O. B., Araújo, A. C. G., Martins, A. C. Q., Saraiva M. A. P., Oliveira, T. N., Togawa, R. C., Leal-Bertioli, S. C. M., Bertioli, D. J., Brasileiro, A. C. M.(2015). Root transcriptome analysis of wild peanut reveals candidate genes for nematode resistance. *PLoS ONE,* 10(10), p.22.

Guimaraes, L. A., Mota, P. A. Z., Araújo, A. G. C., Figueiredo, L. F. A., Pereira, B. M., Saraiva, M. A. P., Silva, R. B., Danchin, E. G. J., Guimaraes, P. M., Brasileiro, A. C. M. (2017).Genome-wide analysis of expansin superfamily in wild *Arachis* discloses a stress-responsive expansin-like B gene. *Plant Molecular Biology.*

Hajdukiewicz, P., Svab, Z., Maliga, P. (1994). The small, versatile pPZP family *oíAgrobacterium* binary vectors for plant transformation. *Plant Mol Biol* 25: 989- 994.

Han, Y., Chen, Y., Yin, S., Zhang, M., Wang, W. (2015). Over-expression of TaEXPB23, a wheat expansin gene, improves oxidative stress tolerance in transgenic tobacco plants. J. *Plant Physiol.,* 173,62-71.

Hildebrand, E.M. (1934). Life History of the hairy-root organism in relation to its pathogenesis on nursey apple trees. *Journal of agricultural research* 48: 857-885 p. 857-885.

Holbrook, C., Stalker, H. (2003). Peanut breeding and genetic resources. Plant Breeding *Reviews,* 22:297-355.

Huang, C. S., Charchar, J. M. (1981) Period of permanence of *Crotalaria spectabilis in* the field influencing the control of meloidoginosis in carrots. *Fitopatol.* Brasília, v.6, supplement, p.538-539.

Huang, C. S., Mota e Silva, E. F. S. (1980). Interruption of the life cycle of *Meloidogyne incognita* by *Crotalaria* spp. *Fitopatol.* Brasília, v.5, supplement, p.402-403.

Hilu, K. W.,Stalker, H. T. (1995). Genetic relationships between peanut and wild species of *Arachis* (Fabaceae): Evidence from RAPDs. *Plant Systematics and Evolution, N.* 198, pp 167-178.

Ibrahim, H., Hosseini, P., Alkharouf., Hussein, E., Gamai El-Din, A.E.K., Aly, M., Matthews, B. (2011). Analysis of Gene expression in soybean *(Glycine max)* roots in response to the root knot nematode Meloidogyne incognita using microrrays and KEGG pathways. BMC *Genomics* 12:220.

Jung, S., Tate, P. L., Hom, R., Kochert, G., Moore, K. and Abbot, A. G. (2003). The Phylogenetic Relationship of Possible Progenitors of the Cultivated Peanut. *Journal of Heredity* 94(4): 334-340.

Kuma, K.M., Lopes-Caitar, V.S., Romero, C.C.T., Silva, S.M.H., Kuwahara, M.K., Carvalho,M.C.C.G., Abdelnoor, R.V., Dias, W.P., Marcelino-Guimaraes, F.C. (2015). A high efficient protocol for soybean root transformation by *Agrobacterium rhizogenesand* most stable reference genes for RT-qPCR analysis. *Plant Cell Rep.,* 34, 1987-2000.

Kereszt, A., Li, D., Indrasumunar, A., Nguyen, C. D., Nontachaiyapoom, S., Kinkema, M., Gresshoff, P. M. (2007). *Agrobacterium rhizogenes-raQàxaXed* transformation of soybean to study root biology. *Nat Protoc.*2:948-952.

Kim, J., Lee, S., Park, S. (2010). Resveratrol production in hairy root culture of peanut, *Arachis hypogaea* L. transformed with different *Agrobacterium rhizogenes* strains.

African JBiotechnol 7:3788-3790.

Kochert, G., Stalker, H. T., Gimenes, M., Galgaro, L., Lopes, C. R. and Moore, K. (1996). RFLP and cytogenetic evidence on the origin and evolution of allotetraploid domesticated peanut *Arachis hypogaea* (Leguminosae). *American Journal of Botany* 83: 11282-11291.

Kosuta, S., Hazledine, S., Sun, J., Miwa, H., Morris, R.J., Downie, J.A., Oldroyd, G.E.D. (2008). Differential and chaotic calcium signatures in the symbiosis signalling pathway of legumes. *Proceedings of the National Academy of Sciences* 105: 9823-9828.

Krapovickas, A., Gregory, W. C. (1994). Taxonomy of the genus *Arachis* (Leguminosae). *Bonplandia,* v.8, p .1-186.

Lacorte, C., Barros, L. M. G.(2015). Reporter genes gus (beta-glucuronidase) and gfp (green-fluorescent protein). In: Brasileiro, A. C. M., Carneiro, V. T. C. (Org.). *Manual de Transformação Genética de Plantas.* 2ed. Brasília: Embrapa, p. 147-164. 2015.

Lavia, G., Femández, A., Simpson C., Seijo, G. (2001). Cytogenetic analysis in wild diploid Arachis species (Leguminosae). *Cytologia* 66: 293-298.

Leal-Bertioli, S. C. M., Guimaraes, P. M., Bruzzi, M., Carneiro, R. G., Valls, J. F.M., Bertioli. D. J. (1999). Search for resistance to the gall nematode *Meloidogyne* spp. and sequences analogous to resistance genes in wild accessions of *Arachis.*
EMBRAPA *Research Bulletin.* N° 20.

Lefebvre, B., Timmers, T., Mbengue, M., Moreau, S., Hervé, C., Tóth, K., Bittencourt-Silvestre, J., Klaus, D., Deslandes, L., Godiard, L., Murray, J.D., Udvardi, M.K., Raffaele, S., Mongrand, S., Cullimore, J., Gamas, P., Niebel, A., Ott, T. (2010). A remorin protein interacts with symbiotic receptors and regulates bacterial infection. *Proceedings of the National Academy of Sciences* 107: 2343- 2348.

Li, J., Todd, T. C., Trick, H. N. (2010). Rapid in planta evaluation of root expressed transgenes in chimeric soybean plants. *Plant Cell* Rep 29:113-123.

Lin, M-H.,Gresshoff, P. M., Indrasumunar, A., Ferguson, B. J. (2011). pHairyRed: A novel binary vector containing the *DsRed2* reporter gene for visual selection of transgenic hairy roots. *Molecular plant, N.* 4, n. 3, p.

Liu, S., Su, L., Liu, S., Zeng, X., Zheng, D., Hong, L., Li, L. *Agrobacterium rhizogenes* -mediated transformation of *Arachis hypogaea:* an efficient tool for functional study of genes. *BiotechnolEquip* 2818:1-10.

Matthews, B.F., Beard, H., MacDonald, M.H., Kabir, S., Youssef, R.M., Hosseini, P., Brewer, E. (2013). Engineered resistance and hypersusceptibility through functional metabolic studies of 100 genes in soybean to its major pathogen, the soybean cyst nematode. *Planta* 237:1337-1357.

Marowa, P., Ding, A., Kong, Y. (2016). Expansins: roles in plant growth and potential applications in crop improvement. *Plant CellB&p.,* 1-17.

McQueen-Mason, S., Durachko, D.M., Cosgrove, D.J. (1992). Two endogenous proteins that induce cell wall extension in plants. *The Plant Cell,* 4, 1425-1433.

Medina-Bolivar, F., Condori, J., Rimando, A.M., Hubstenberger, J., Shelton, K., O'Keefe, S.F., Bennett, S., Dolan, M.C. (2007). Production and secretion of resveratrol in hairy root cultures of peanut. *Phytochemistry* 68:1992-2003.

Mehrotra, S., Srivastava, V., Rahman, LU., Kukreja, a. K. (2015).Hairy root biotechnology - indicative timeline to understand missing links and future outlook. *Protoplasma* 252:1189-1201.

Miller, J. H. (1972). Experiments in molecular genetics. New York: Cold Spring Harbor 466 pp.

Mitreva, M., Blaxter, L. M. R., Bird, D. M., Cárter, J. P. (2005). Comparative genomics of nematodes. *Trends in Genetics* 21(10):573-581.

Moraes, S.A., Salgado, C.L. (1982). Use of the technique of detached leaves of peanut *(Arachis hypogaea* L.) for inoculations with *Cercospora arachidicola* Hori and *Cercospora personata* (Bert. e Curt.) Eli. e Ev. *Summa Phytopathologica,* v.8, p.39-55.

Morgante, C. V., Brasileiro, A. C. M., Roberts, P. A., Guimaraes, L. A., Araújo, A. C. G., Fonseca, L. N., Leal-Bertioli S., Bertioli, D. J., Guimaraes, P. M. (2013).A survey of genes involved in *Arachis stenosperma* resistance to *Meloidogyne arenaria* race 1. *Functional Plant Biology.* 40 (12): 1298-309.

Narayanan, R. A., Atz, R., Denny, R., Young, N. D., Somers, D. A. (1999). Expression of Soybean Cyst Nematode Resistance in Transgenic Hairy Roots of Soybean Contribution no. 991130114 from the Minnesota Agric. Exp. Stn. This work was supported in part by USDA/95-37300- 1593. *Crop Sei,* 39:1680-1686.

Novak, A., Guo, C., Yang, W., Nagy, A., Lobe, C.G. (2000). Z/EG, a double reporter mouse 51 line that expresses enhanced green fluorescent protein upon Cre-mediated excision. *Genesis* 28: 147-155.

Pak, H.K., Sim, J.S., Rhee, Y., Ko, H.R., Ha, S.H., Yoon, M.S., Kang, C.H., Lee, S., Kim, Y.H., Hahn, B.S. (2009). Hairy root induction in Oriental melon *(Cucumis melo}* by *Agrobacterium rhizogenes* and reproduction of the root-knot nematode *(Meloidogyne incognita}. Plant Cell Tissue Organ Cult* 98:219-228.

Plovie, E., De Buck, S., Goeleven, E., Tanghe, M., Vercauteren, L, Gheysen, G. (2003). Hairy roots to test for transgenic nematode resistance: *Think twice. Nematology* 5:831-841.

Proite, K., Leal-Bertioli, S., Bertioli, D., Moretzsohn, M., Silva, F., Martins, N., Guimaraes, P. (2007). ESTs from a wild *Arachis* species for gene discovery and marker development. BMC *Plant Biol.* 7:7.

Proite, K., Carneiro, R., Falcão, R., Gomes, A., Leal-Bertioli, S., Guimaraes, P., Bertioli, D. (2008).Post-infection development and histopathology of *Meloidogyne arenaria* race 1 *on Arachis* spp. *Plant Pathology,* 57 (5):974-80.

Riker, A. J., Banfield, W. M., Wright, W. H., Keitt, G. W., Sagen, H. E. (1930). Studies on infectious hairy root of nursey Apple trees. *Journal of Agricultural Research,* v. 41 No. 7 pp. 507-540pp.

Seijo, J. G., Lavia, G. L, Femández, A., Krapovickas, A., Ducasse, D. and Moscone, E. A. (2004). Physical mapping of the 5S and 18S-25S rRNA genes by FISH as evidence that *Arachis duranensis* and *Arachis ipaensis* (Leguminosae). *American Journal of Botany* 91: 1294-1303.

Semblat, J. P., Bongiovanni, M., Wajnberg, E., Dalmasso, A., Abad, P., Castagnone-Sereno, P. (2000). Virulence and molecular diversity of parthenogenetic root-knot nematodes, *Meloidogyne* spp. *Heredity.* 84: 81-89.

Simpson, R.B., Spielmarm, A., Margossian, L., McKnight, T.D.(1986). A disarmed binary vector from *Agrobacterium tumefaciens* functions in *Agrobacterium rhizogenes'.* frequent co-transformation of two distinct T-DNAs. *Plant Mol. Biol.* 6, 403-415.

Sonia, Saini, R., Singh, R.P., Jaiwal, P. K. *Agrobacterium tumefaciens* mediated transfer of *Phaseolus vulgaris* a-amylase inhibitor-1 gene into mungbean *Vigna radiata* (L.) Wilczek using *bar* as selectable marker. *Plant Cell* Rep 26:187- 198.

Taylor, A. L., Sasser, J. N. (1978). Biology, identification and control of root-knot nematodes (Meloidogyne species). NCSU and USAID Cooperative Publication, Raleigh, NC, U.S.A.

Tillman, B., Stalker, H. T. (2010) Peanut. In: VollmannJ, RajcanI, editors. Oil Crops:

Springer New York; 2010. p. 287-315.

Tirichine, L., Sandal, N., Madsen, L.H., Radutoiu, S., Albrektsen, A.S., Sato, S., Asamizu, E., Tabata, S., Stougaard, J. (2007). A gain-of-function mutation in a cytokinin receptor triggers spontaneous root nodule organogenesis. *Science* 315: 104107.

Trudgill, D. L., Blok, V. C. (2001). Apomitic, polyphagous root-knot nematodes: exceptionally successful and damaging biotrophic root pathogens. *Annual Review of Phytopathology.* 39: 53-77.

Tzortzakakis, E. A., Gowen, S. R., Goumas, D. E. (1996) Decreased ability of *Pasteuria penetrans* spores to attach to successive generations of *Meloidigyne javanica. FundAppl Nematol* 19:201-204.

USDA-FAS. 2015. *Table 01: Major Oilseeds: World Supply and Distribution (Commodity View).* Accessed in December 2016, available at USDA - United States Department of Agriculture/FAS - Foreign Agricultural Service: https://apps.fas.usda.gov/psdonline/circulars/oilseeds.pdf

Valls, J. F. M., Simpson, C. E., (1994). Taxonomy, natural distribution, and attributes of *Arachis.* In: Kerridge, P.C.; Hardy, B. (EDs.). Biology and Agronomy of Forage *Arachis.* Cali: CIAT, p.1-18.

Valls, J. F. M., Simpson C. E. (2005) New species of *Arachis* L. (Leguminosae) from Brazil, Paraguay and Bolivia. *Bonplandia* 14:35-64.

Vargas Gil, S., Haro, R., Oddino, C., Keamey, M., Zuza, M., Marinelli, A., March, G.J. (2008). Crop management practices in the control of peanut diseases caused by soilbome fungi. *Crop Protection* 27: 1-9.

Yang, H., Naim, J., and Ozias-Akins, P. (2003). Transformation of peanut using a modified bacterial mercuric ion reductase gene driven by an actin promoter from *Arabidopsis thaliana. J. Plant Physiol.* 160:945-952.

White, L. O. (1972). The taxonomy of the srow gall organism *Agrobacterium tumefaciens* and its relationship to Rhizobia and to other *Agrobacterium. J. Gen Microbiol* 77:565-574.

Williamson, V. M., Gleason, C. A. (2003). Plant-nematode interactions. Current Opinion in *Plant Biology* 6:327-33.

Williamson, V. M.,Hussey, H. R. (1996). Nematode pathogenesis and resistance in plants. *The Plant Cell* 8: 1735 - 1745.

Wynne, J. C., Beute, M. K., Nigam, S. N. (1991). Breeding for disease resistance in peanut *(Arachis hypogaea* L.). *Annual Review of Phytopathology* 29:279-303.

Zinovieva, S. V., Vasyukova, N. L, Udalova, Z. V., Gerasimova, N. G.,Ozeretskovskaya, O. L. (2011). Involvement of salicylicacid in induction of nematode resistance in plants. *Biol Buli Russ AcadSei*, 38(5):453-8.

CHAPTER 9

ANNEXES

Table 1 - Evaluation of the number of galls and egg masses per gram of transgenic root on a detached peanut leaf, 90 days after inoculation with *M. arenaria*.

Transgenic event (root)*	Weight (g)	Total number of galls	Total number of egg masses	Number of galls/g	Number of egg masses/g
pPZP-1	0,06	1	1	16,67	16,67
pPZP-2	0,18	4	3	22,22	16,67
pPZP-3	0,21	2	2	9,52	9,52
pPZP-4	0,22	4	3	18,18	13,64
pPZP-5	0,30	8	4	26,67	13,33
pPZP-6	0,25	5	2	20,00	8,00
pPZP-7	0,30	6	3	20,00	10,00
pPZP-8	0,21	5	3	23,81	14,29
pPZP-9	0,34	4	4	11,76	11,76
pPZP-10	0,28	6	5	21,43	17,86
Average	**0.24**	**4,50**	**3,00**	**19,03**	**13,17**
pPZP-AdEXLB8-1	0,24	0	0	0	0
pPZP-AdEXLB8-2	0,16	0	0	0	0
pPZP-AdEXLB8-3	0,13	0	0	0	0
pPZP-AdEXLB8-4	0,21	0	0	0	0
pPZP-AdEXLB8-5	0,28	1	0	3,57	0
pPZP-AdEXLB8-6	0,08	0	0	0	0
pPZP-AdEXLB8-7	0,36	0	0	0	0

pPZP-AdEXLB8-8	0,39	0	0	0	0
pPZP-AdEXLB8-9	0,24	0	0	0	0
pPZP-AdEXLB8-10	0,34	1	1	2,94	2,94
pPZP-AdEXLB8-11	0,24	0	0	0	0
pPZP-AdEXLB8-12	0,34	0	0	0	0
pPZP-AdEXLB8-13	0,17	0	0	0	0
pPZP-AdEXLB8-14	0,12	0	0	0	0
Averages	**0.24**	**0,14**	**0,07**	**0,47**	**0,21**
pPZP-AsDUF-1	0,20	1	1	5,00	5,00
pPZP-AsDUF-2	0,35	2	1	5,71	2,86
pPZP-AsDUF-3	0,15	0	0	0,00	0,00
pPZP-AsDUF-4	0,30	1	1	3,33	3,33
pPZP-AsDUF-5	0,33	1	1	3,03	3,03
pPZP-AsDUF-6	0,20	0	0	0,00	0,00
pPZP-AsDUF-7	0,13	0	0	0,00	0,00
pPZP-AsDUF-8	0,45	2	2	4,44	4,44
pPZP-AsDUF-9	0,35	1	1	2,86	2,86
pPZP-AsDUF-10	0,30	0	0	0,00	0,00
Averages	**0,28**	**0,80**	**0,70**	**2,44**	**2,15**
pPZP-AsLIP-1	0,34	3	2	8,82	5,88

pPZP-AsLIP-2	0,38	3	2	7,89	5,26
pPZP-AsLIP-3	0,27	2	1	7,41	3,70
pPZP-AsLIP-4	0,25	1	1	4,00	4,00
pPZP-AsLIP-5	0,28	3	2	10,71	7,14
pPZP-AsLIP-6	0,20	2	1	10,00	5,00
pPZP-AsLIP-8	0,30	3	2	10,00	6,67
pPZP-AsLIP-9	0,31	2	2	6,45	6,45
pPZP-AsLIP-10	0,32	2	3	6,25	9,38
pPZP-AsLIP-11	0,22	2	2	9,09	9,09
pPZP-AsLIP-12	0,14	1	1	7,14	7,14
Averages	**0,27**	**2,17**	**1,67**	**8,04**	**6,17**

*In this study, the set of roots emitted from the inoculation site was analysed as a transgenic event.

Table 2 - Evaluation of the numbers of galls and egg masses per gram of transgenic root in composite soya plants, 90 days after inoculation with *M. incognita*.

Transgenic event (root)*	Weight (g)	Total number of galls	Total number of egg masses	Number of galls/g	Number of egg masses/g
pPZP-1	1,25	35	30	28,00	24,00
pPZP-2	3,78	61	55	16,14	14,55
pPZP-3	3,05	49	41	16,07	13,44
pPZP-4	3,30	45	44	13,64	13,33
pPZP-5	2,08	72	65	34,62	31,25
pPZP-6	2,85	35	31	12,28	10,88
pPZP-7	1,89	25	23	13,23	12,17
pPZP-8	3,39	35	34	10,32	10,03
pPZP-9	1,25	29	27	23,20	21,60

pPZP-10	4,48	61	60	13,62	13,39
pPZP-11	3,62	49	47	13,54	12,98
pPZP-12	4,23	58	59	13,71	13,95
pPZP-13	2,25	36	38	16,00	16,89
pPZP-14	4,51	57	57	12,64	12,64
pPZP-15	3,81	52	53	13,65	13,91
pPZP-16	2,12	33	34	15,57	16,04
pPZP-17	3,31	46	47	13,90	14,20
pPZP-18	3,23	41	41	12,69	12,69
pPZP-19	2,25	31	32	13,78	14,22
pPZP-20	1,25	35	30	28,00	24,00
Average	**2,98**	**44,74**	**43,05**	**16,09**	**15,42**
pPZP-AdEXLB8-1	6,01	5	5	0,83	0,83
pPZP-AdEXLB8-2	5,19	4	4	0,77	0,77
pPZP-AdEXLB8-3	4,22	3	3	0,71	0,71
pPZP-AdEXLB8-4	4,43	4	3	0,90	0,68
pPZP-AdEXLB8-5	3,06	2	2	0,65	0,65
pPZP-AdEXLB8-6	5,08	4	4	0,79	0,79
pPZP-AdEXLB8-7	2,44	0	0	0,00	0,00
pPZP-AdEXLB8-8	2,83	2	2	0,71	0,71
pPZP-AdEXLB8-9	3,73	2	2	0,54	0,54
pPZP-AdEXLB8-10	4,51	4	3	0,89	0,67
pPZP-AdEXLB8-11	3,62	3	3	0,83	0,83

pPZP-AdEXLB8-12	3,31	3	2	0,91	0,60
pPZP-AdEXLB8-13	4,96	4	5	0,81	1,01
pPZP-AdEXLB8-14	2,09	0	0	0,00	0,00
pPZP-AdEXLB8-15	3,39	0	0	0,00	0,00
pPZP-AdEXLB8-16	2,24	3	4	1,34	1,79
pPZP-AdEXLB8-17	2,34	2	2	0,85	0,85
pPZP-AdEXLB8-18	4,19	2	3	0,48	0,72
pPZP-AdEXLB8-19	2,09	1	1	0,48	0,48
pPZP-AdEXLB8-20	3,16	2	2	0,63	0,63
pPZP-AdEXLB8-21	2,74	3	2	1,09	0,73
Averages	**3,60**	**3**	**2**	**0,68**	**0,67**
pPZP-AsDUF-1	5,19	13	13	2,50	2,50
pPZP-AsDUF-2	5,15	12	12	2,33	2,33
pPZP-AsDUF-3	5,76	14	13	2,43	2,26
pPZP-AsDUF-4	4,82	12	12	2,49	2,49
pPZP-AsDUF-5	3,89	10	10	2,57	2,57
pPZP-AsDUF-6	2,15	7	7	3,26	3,26
pPZP-AsDUF-7	3,41	12	11	3,52	3,23

pPZP-AsDUF-8	5,65	10	10	1,77	1,77
pPZP-AsDUF-9	0,89	0	0	0,00	0,00
pPZP-AsDUF-10	3,15	9	9	2,86	3,49
pPZP-AsDUF-11	1,64	3	4	1,83	2,44
pPZP-AsDUF-12	3,86	8	10	2,07	2,59
pPZP-AsDUF-13	2,68	9	9	3,36	3,36
pPZP-AsDUF-14	3,02	5	6	1,66	1,99
pPZP-AsDUF-15	3,16	10	11	3,16	3,48
Averages	**3,62**	**9**	**9**	**2,38**	**2,52**

*In this study, the set of roots emitted from the inoculation site was analysed as a transgenic event.

FSC
www.fsc.org
MIX
Papier aus verantwortungsvollen Quellen
Paper from responsible sources
FSC® C105338

Printed by Books on Demand GmbH, Norderstedt / Germany